Other books by
Stephen Budiansky

The Nature of Horses

Covenant of the Wild

Nature's Keepers

IF A LION ᴋᴋ ᴋᴋ
COULD TALK

ANIMAL INTELLIGENCE AND THE
EVOLUTION OF CONSCIOUSNESS

STEPHEN BUDIANSKY

THE FREE PRESS

New York   London   Toronto

Sydney   Singapore

THE FREE PRESS
A Division of Simon & Schuster Inc.
1230 Avenue of the Americas
New York, NY 10020

Designed by Carla Bolte

Manufactured in the United States of America

10   9   8   7   6   5   4   3   2   1

Library of Congress Cataloging-in-Publication Data

Budiansky, Stephen.
     If a lion could talk : animal intelligence and the evolution of
consciousness / Stephen Budiansky.
        p.   cm.
     Includes bibliographical references (p.   ) and index.
     1. Animal intelligence.   2. Consciousness in animals.   I. Title.
     QL785.B823 1998
     591 . 5' 13—dc21   98-28211
                                                                   CIP

     ISBN 978-1-5011-4274-1

For my parents

If a lion could talk,

we would not understand him.

—*Ludwig Wittgenstein*

CONTENTS

ACKNOWLEDGMENTS

I WOULD LIKE TO EXPRESS MY DEEP APPRECIATION TO the many scientists who assisted me in my research for this book. I am especially grateful to Jacques Vauclair and Euan Macphail, who read draft chapters and offered invaluable criticisms and suggestions.

The National Library of Medicine in Bethesda, Maryland, is a priceless resource for anyone researching a topic such as this, and to those who had the wisdom to create that institution and open its doors to all comers, I give my heartfelt thanks.

INTRODUCTION

GORILLA SAVES TOT IN BROOKFIELD ZOO APE PIT
ran the page-one headline of the *Chicago Tribune*. The story, quickly
picked up around the world as television crews from Britain, Germany,
Australia, and the ever-present CNN descended on the scene, was as
follows: Binti, an eight-year-old female lowland gorilla and "bona fide
hero," had rescued a three-year-old boy who had climbed over a railing
and fallen into an enclosure with seven gorillas. The reports told how
Binti picked up the unconscious boy, cradled him in her arms, and carried him gently to a door to the apes' enclosure where paramedics were
waiting. "Another gorilla walked toward the boy, and she kind of
turned around and walked away from the other gorillas and tried to be
protective," one visitor told reporters. A keeper at another zoo who had
helped raise Binti during the first month of the gorilla's life and who
had watched the rescue on television said, "I could not believe how gen-
tle she was. I just had chills."

Visitors who had read the story poured into the zoo to see Binti.
Some told reporters that they had cried when they heard about the
ape's marvelous feat. Others sent gift bananas or called to sign up for
the zoo's "adoption" program—a $25 or greater annual contribution
to help pay for the care and feeding of the zoo animal of your choice.

Binti's saga is the prototypical animal story of our age. Stories of elephants that grieve and elephants that paint pictures, of dolphins that rescue swimmers in distress, of talking parrots that say "I love you! I'm sorry!," of gorillas that insist on watching Pavarotti on TV, of dogs that pull their owners from burning buildings—such tales are seized on with an eagerness that transcends our mere tendency toward the sentimental. The alleged sagacity, insight, even morality or compassion of animals is taken as proof that animals do not merely *know*, but know *better* than selfish and brutal humankind.

Binti's heroic rescue is typical in another way: It is not exactly true. The widely circulated version of events omitted several crucial details. For one thing, Binti did not in fact shield the injured boy from the other gorillas. That job fell to the zoo keepers, who quickly responded (like good Southern sheriffs) by turning on three high-pressure fire hoses and training them at the other gorillas' feet. By these means the other gorillas were shooed out another door of their pen. "The media made it sound like Binti made a conscious decision to, quote unquote 'save the boy,' but this is speculation," lead keeper Craig Demitros said six months later at a workshop where zoo keepers met to discuss maternal behavior in gorillas. "She saved him from what, really? The other animals were not coming after him," he said.

The other missing detail in the story was even more telling. Many gorillas reared in zoos do not develop proper maternal behavior—which in mammals often involves not only instinct but learning. Binti, herself neglected by her mother and hand-reared by humans, received extensive training in how to be a mother when she became pregnant. The zoo keepers trained her to carry a doll and, moreover, to retrieve it and bring it to her keepers. In other words, Binti was just doing what she had been trained to do—no different from the way a search-and-rescue dog is trained to track a human's scent, or, for that matter, the way a retriever is trained to fetch a stick. The fact that the boy had been stunned by his fall of more than twenty feet helped, too. "If he had been awake and screaming, he might well have elicited a different kind of response," Demitros explained—Binti might have run away, or even pushed or bitten the boy.

Those who seize on such tales as proof that animals are far more remarkable than science has given them credit for have, ironically, adopted a profoundly self-centered definition of intelligence. What invariably provokes comment is how closely the animal seems to have resembled man's capacities for thought and emotion. Tales of elephants grieving for their dead, of red foxes as doting fathers full of parental love for their cubs, of chimpanzees that silently watch sunsets—such are the stories that animal rightists invoke in their effort to knock man off his anthropocentric pedestal at the top of creation. What makes an animal worthy of special consideration, they are effectively saying, is how closely its behavior resembles that of a human (or at least a human on a good day). In their battle against anthropocentrism they have adopted the most anthropocentric stance imaginable. It is an argument as curious as it is revealing.

If the modern sciences of evolutionary biology and ecology have taught us anything, it is that life generates diversity; the millions of species on earth each reflect millions of years of separate adaptation to unique environments and unique ways of life. The mind is no exception to the facts of natural selection; it makes as little sense to expect that other species should share the uniquely human thought processes of the human mind as it would to expect that we should share an elephant's trunk or a zebra's stripes. And as we shall see, evolutionary ecology, the study of how natural selection has equipped animals to lead the lives they do, is beginning to tell us much about how the minds of animals process information in ways that are uniquely their own.

CONTINUITY AND DUALISM

How animals think, and what they think about, are ancient questions that have proved both irresistible and maddeningly elusive. Historically there have been two great currents in human thought on the subject. One emphasizes the continuity between man and other animals; the other emphasizes the discontinuities, or the duality, that separate man and the rest of creation. Both views have deep roots. Traditional tales of the North American Indians and of African tribes often eradicated the discontinuity altogether with accounts of animals that changed to humans and humans that changed to animals. The ancient Greeks, not only in their fables but in their serious philosophical ruminations, credited animals (the scheming fox, for instance) with humanlike motives and insight.

Stories of benevolent animals that warn their masters of danger appear in the folkways of virtually every human culture. The added twist in many of these stories is that the master refuses to understand what the animal has done, and in anger and misunderstanding kills the animal or allows it to die, only later to discover the truth. In *A Thousand and One Nights* we hear of a falcon that saved the life of the king and his horse: The king, tired from a hot day's hunt, takes a bowl carried about his falcon's neck and fills it with water he sees running down the trunk of a tree. The falcon repeatedly knocks the bowl over before the horse can drink. The king becomes angered and slices off the bird's wings with his sword, whereupon the bird lifts its head "as if to say, 'Look into the tree!'" Only then does the king see that what he took to be water running down the tree was venom being spit by an enormous snake. Essentially the same tale, with a shepherd's dog in the starring role, appears in one of Rabbi Meir's fables in the Talmud, the compilation of Jewish biblical commentaries dating from the first century. In this version the dog actually drinks the poison himself to save his master, who has ignored the dog's frantic attempts to warn him.

Traditional Western religion, however, has perhaps more frequently adopted a dualist stance. Man was instructed by no less an authority than God himself to subdue the earth, and Judaism's emphasis on man as an ethical agent, and Christianity's emphasis on the human soul (not to mention medieval Christianity's abiding distrust of vestiges of pagan animal worship), drew as sharp a line as one could possibly draw between man and non-man. Recent surveys have found that fundamentalist Christians remain among those most likely to strongly reject the notion of emotional or intellectual continuity between man and animals.

They are also the most likely to reject evolution. Darwin's theory is the scientific touchstone for champions of continuity. If man and animals have a common ancestry, if man was not the product of special creation in the image of God, then why should we not expect animals to share our fundamental capacity for rational choice, sense, reflection, insight, and feeling? Darwin's conclusion in *The Descent of Man* that "the difference in mind between man and the higher animal, great as it is, is certainly one of degree and not of kind"—that "the senses and intuitions, the various emotions and faculties, such as love, memory, attention, curiosity, imitation, reason, etc., of which man boasts, may be found in an incipient or even sometimes in a well-developed condition, in lower animals"—is religiously quoted by those scientists who would press to the utmost the implications of ape-language studies and anecdotes of animal creativity, insight, and deception.

But many who cite Darwin to this effect leave out Darwin's parenthetical acknowledgment of enormous practical difference between human and animal minds ("great as it is"). And elsewhere in *The Descent of Man* Darwin observed that "a moral being is one who is capable of reflecting on his past actions and their motives—of approving some and disapproving of others; and the fact that man is the one being who certainly deserves this designation is the greatest of all distinctions between him and the lower animals." Aristotle called man the sole "political animal"; Darwin said man was the only moral animal or "utopian animal," in that he could modify his actions in obedience to an ideal that exists beyond any immediate, real experience.

ANTHROPOMORPHISM AND DOGOMORPHISM

This very tension between the sameness and otherness of animals may explain why we want so much to see into their mental lives. They are similar to us in so many ways and yet so different. Animals, especially the animals that so many of us share our houses and lives with, act and react in ways we find so familiar, yet they are a tantalizing closed book. The gift of human language allows us to penetrate the interior lives of our fellow men; we can read Montaigne's essays and glimpse what it was like to be a sixteenth-century French nobleman or talk to a computer programmer from Berkeley or a lama from Tibet and ask as many questions as we like about their lives, their thoughts, their feelings, their desires, their hopes, their pasts, and their futures. We do communicate with animals, but never in a way that permits them to describe their experience of being.

If our greatest obstacle to entering the mind of another animal is its inability to communicate as we do, the second greatest is our self-centered way of looking at the world. It is a fault we share with other species. It is commonplace to speak of a dog that thinks it is human, but a better statement of the situation is that the dog thinks we're dogs. Funny looking dogs, to be sure, and dogs that refuse to engage in the full array of normal dog behavior, but dogs that are enough like dogs to get along with under the working assumption they are dogs. Our dogs sniff us as they sniff other dogs, bow to us with outstretched front legs when they want to play just as they do to other dogs, and perhaps most important, submit to us as they submit to a pack leader. Cats are descendants of far less social creatures, but when they notice us at all, we're cats. They deliver dead mice to us, or beg for attention the way a kitten does from its mother with its tail straight up in the air. They paint the world in their image just as we paint it in ours. "If cattle and horses, or lions, had hands, or were able to draw with their feet and produce the works which men do," wrote Xenophanes 2,500 years ago, "horses would draw the forms of gods like horses, and cattle like cattle, and they would make the gods' bodies the same shape as their own."

So is it only natural for us to see a dog's behavior in the terms *we* know. When a dog defecates on the oriental rug and greets us at the door cringing, we without hesitation say he is feeling guilty over what he has done. When a horse nuzzles us, we say he likes (or even loves) us. If we are particularly competitive we say our dog or our horse loves to win blue ribbons at shows. If we are particularly given to New Age mystification, we can suggest the following explanations for why a mother lion ate her dead cub after it was killed by a male from another pride: "Maybe she felt closer to her dead offspring when it was part of her body once again. Maybe she hates waste, or cleans up all messes her cubs make, as part of her love. Maybe this is a lion funeral rite." Or maybe she was hungry.

Anthropomorphism, the tendency to view an animal's actions in terms of our own conscious intentions, thoughts, and motives, is viewed by many as an act of generosity toward the species we aim it at, and humility on our part. The ultimate compliment: You're almost human! But a more honest evaluation may be that anthropomorphism betrays an utter lack of imagination on our part—not to mention a slavish obedience to an instinct that may well have been pounded into our genes over the course of millions of years of evolution. The eminent animal behaviorist John S. Kennedy, formerly of the University of London, has labeled our behavior "compulsive anthropomorphism," so irresistible does this tendency seem to be. Our very language almost compels us to describe phenomena in anthropomorphic terms, terms that ascribe intention and purpose even to inanimate objects. A particular plant "likes" shade. A malfunctioning engine is "giving us trouble." A computer is "trying to figure out" which printer to send a file to.

Students of evolution have long wrestled with the problem of how to avoid language that implies a purpose or goal, and by all appearances have just about given up. The "selfish gene" metaphor has wide currency: more exacting descriptions are necessarily long-winded and cumbersome. We are forever saying things like, "as the forests gave way to open grassland, horses evolved longer legs in order to run fast and escape from predators." Of course horses did nothing of the sort— they never evolved anything "in order" to do anything. What happened was that those individual horses who, through chance mutation and re-combination of genes, had longer legs were more likely to survive and pass on those genes to their offspring. But the very nature of communication always favors the succinct; there is essentially no other way of conveying the thought quickly than by invoking the metaphor of intentionality.

Man's readiness to ascribe human motives and intentions to phenomena at large is manifest in primitive peoples' view of violent weather or volcanic eruptions or earthquakes as a "punishment"; in the classical personification of Fate and Fortune and Love; and indeed in man's almost obsessive search for meaning in everyday life ("why did this tire have to go flat today!"). The most sweeping example is the nearly universal urge to believe that behind the otherwise inexplicable workings of probability is an all-knowing God who has a reason for everything that he chooses to happen to us.

Natural selection may have favored our tendency to anthropomorphize: Being able to guess at the motives of our fellow man has clear survival value. Being good at thinking "what would I do in his position" can help us calculate what our rivals may be up to and outsmart them. It can help us avoid conflict by anticipating trouble. Because evolution's relentless selection for adaptive traits has so often honed in animals instinctive behaviors that bear a strong working resemblance to the action of purposeful intent, our tendency to anthropomorphize the animals we hunt may have given us a huge advantage in anticipating their habits and their evasions. But it also has made us very bad at being objective about the true nature of the things in the world that actually are not like us.

So we are predisposed—if not preprogrammed—to accept tales of animals who display human motives, understanding, reason, and intentions. It takes a far greater imagination to conceive the possibility that a dog's mental life may assume a form that is simply beyond our ken.

In fact the most astonishing things (astonishing to us, that is) that animals do almost certainly have nothing whatever to do with conscious thought as *we* know it. A horse's or a pigeon's or a bee's ability to find its way to food or home; a monarch butterfly's ability to migrate thousands of miles with unerring accuracy; a chimney swift that catches insects with routine precision in midflight; a sheepdog that strikes the precise balance between ineffectuality and chaos to push a flock of sheep forward without them either stopping or scattering; a dressage horse that executes dozens of all but imperceptible commands—these, after all, are the things that animals do superbly well every day of their lives. We are drawn to the coincidental oddities of animals that mimic a human mental proficiency, yet tend to ignore the prodigious feats that animal minds routinely pull off right before our eyes.

KNOWING THE OTHER

To understand what we truly can about how animal minds work inescapably means to abandon any real hope of penetrating their thoughts, or of translating their thoughts into human terms. Perhaps the oldest problem in philosophy is the question of how we can even know if mental experiences exist at all outside of ourselves. For all that any of us know, we are each the only beings in the universe capable of thinking thoughts. For how could we ever have proof to the contrary? Suppose someone demanded of you proof that you are conscious— what could you say to convince him? No matter what your answer, he could insist that you're just a well-programmed computer, that nothing you say or do proves that there is a consciousness within. You might show him a copy of your brain scan, but he could counter that a computer has all sorts of electrical activity going inside it, too; so what?

The same problem, as it applies to computers, has tied philosophers and artificial-intelligence researchers in knots for decades. Suppose you could build a computer that completely simulated a human brain. Would it actually be conscious? Would it *experience* thoughts and feelings as we do? In 1950, the mathematician Alan Turing proposed a simple test: put a computer in one room, a person in the other, and ask each of them questions without peeking. If you can't tell by their answers which is which, the computer is conscious.

In our hearts we all feel there's something wrong with that definition—that doing is not the same as understanding or consciously experiencing. I have a program on my computer called "The Talking Moose." If I haven't typed anything for ten minutes, the moose appears in the upper left-hand corner of the screen and spouts a pithy comment on life. The nice feature is that you can add your own aphorisms to his stockpile of sayings. If you just type words into its catalogue in plain English, the moose does a rather bad job at reading them; he mispronounces, puts the emphasis at the wrong places, ends his sentences dangling in the air. But a few minutes' tweaking with special code characters that control vowel sounds and intonation produces wonders; the moose can read even quite long sentences exactly the way someone who fully comprehended their meanings would. Yet no one could possibly suggest that the moose has even the most elementary understanding of the words he reads.

The philosopher John Searle made a similar point in a powerful attack he launched on the Turing test notion: suppose you put a fluent Chinese speaker in one room and in the other room you put someone who knows no Chinese but is given books listing a complete set of rules for translating and producing Chinese characters and syntax. Now send them both questions in Chinese. The results would be indistinguishable, yet no one could claim that the person following the rules has a conscious understanding of Chinese. Yet, on the other hand, this in a way was Turing's whole point: if you can't tell by outward appearances, how can you ever know?

Or maybe even *our* consciousness is an illusion, too; maybe everything we seem to be experiencing and thinking is not real at all; maybe we're sitting in a sensory deprivation chamber with our neurons plugged into a computer that fires off signals that create the illusion of conscious thought and experience. This is not as fanciful a point as it seems. In one experiment monkeys were first taught to recognize and respond to motion that crosses the visual field in a particular direction. When researchers then electrically stimulated a particular spot of the brain that is associated with motion sensing, the monkeys responded exactly the same as if they had actually seen the real thing.

So maybe this is your fate, too. How would you know? The experimenters get to decide each day what the computer will feed into your brain; when they're in a benevolent mood they arrange for you to think you're having a nice dinner with a $90 bottle of wine and someone else picking up the tab; when they're feeling mischievous they feed you the sensation that you're about to have to take a final exam in that German course you were sure you had dropped at the beginning of the semester.

But behind these well-worn philosophical conundrums is an arguably even greater problem we face when we try to get inside an animal's mind. Simply stated: To truly know what a horse thinks, we'd have to *be* a horse—and then we would have no way of describing horse thoughts in any way recognizable to humans. Not just because horses don't speak or write books—but because we lack any way of expressing or even comprehending what a nonverbal thought experience *is* without falling back on words. Think about how hard it is to describe to someone else (or even to ourselves) the countless mental processes that our brains busy themselves with. Some are completely opaque to us. We are completely unaware of the nerve impulses that keep our heart beating. Only with a great effort are we aware of the link between our minds and the motions of our limbs. You can say to yourself, "now I'm going to lift my arm," but thinking that conscious thought is not what makes your arm go up—try it and you'll find that you really cannot pin down what *thing* it is inside your head that is actually the thing that makes your arm move. Shooting a basketball while running is an extraordinarily complex computational problem that involves coordinating inputs from the eyes and legs, and outputs to the hand and arms, with a vast stored knowledge about expected trajectories; but who can describe what is going on in their minds in the process? We have all had the experience of having a thought and *then* struggling to find the words to express it; yet we find it nigh on impossible to describe *what* that preliterate thought "looked" or "felt" like. Even when we are speaking fluently, we have the sensation (*if* we think about it at all, which we usually do not!) of the words welling up preformed from *somewhere* in our brains. Not even the most ardent animal rights proponent suggests that horses or other nonhuman animals have a secret language of their own that allows them to form concepts into words. To literally see inside a horse's brain would be to enter a world that is without the words to describe it—and so is meaningless to us.

The philosopher Ludwig Wittgenstein made the famous observation, "If a lion could talk, we would not understand him." But that begs the question: if a lion could talk, we probably could understand him. He just would not be a lion any more; or rather, his mind would no longer be a lion's mind.

THE MIDDLE GROUND OF COGNITION

It is conceivable that some day we may be able to decode enough of the "software" of the mind, piecing together clues from which neurons fire under which particular circumstances and which other neurons they connect to, to have a pretty good explanation of consciousness. But that surely will not happen any time soon. Imagine trying to figure out how The Talking Moose works by measuring the voltages in each bit of your computer's memory and processor while the moose talks. Add to the conditions of this problem that you have no clue about the machine language and logical architecture of your (or anyone else's) computer. Good luck.

It might sound like there is thus nothing left to say for the present. But there is a wonderful amount to say about animal minds, so long as we approach the matter with realism and restraint—and respect for what animals do particularly well in the first place, rather than with a determination to make them into slightly defective versions of ourselves. The so-called cognitive revolution in psychology in the 1970s proposed that there is a great deal that animals *do* that perforce requires them to hold mental representations of information and to manipulate those representations, and that appropriately designed experiments can reveal something about what sort of mental *events* may be taking place in an animal's head—even if we can never get at mental *experiences.* This view was a mighty leap from the strict behaviorist position championed by psychologists like B. F. Skinner, who insisted that all behavior was a simple, learned stimulus-response link that told nothing about the inner workings of mind—a position that has been much vilified for reducing animals to automatons. The cognitive approach, while still drawing on an animal's manifest behavior as the chief evidence, relies on experiments that force an animal to use information not in its immediate environment to produce a choice of behaviors.

One class of fascinating experiments, for example, tests animals' abilities to form mental categories. In a typical experiment, capuchin monkeys were shown slides of persons and of "nonpersons," and were rewarded for pressing the correct button that corresponds to each category. The monkeys were then presented with a new set of person and nonperson slides to see how well they had formed a mental concept of the two categories. Such probes of an animal's mind can tell us volumes about its mental capabilities, and even give us insight into how an animal perceives its world—at least as a practical, functional matter—without drifting into the unknowable territory of what it *experiences* as it does so.

Of course realism and restraint is not what everyone is after. Many people are so emotionally committed to a belief in connecting with the mental lives of animals that they refuse to be deflected by any rational argument or prudent skepticism, or to be satisfied with anything science is likely to be able to tell us. Conjectures about the profound truths animals are trying to communicate to us about the meaning of life abound. Members of a new profession called "animal communicators" even offer to tune in to your pet's mind by telepathy and to reveal what they really think of you. Athena, a housecat, for example, reports that she loves her owner but wants to move out because of three new cats that have moved in. "She's in a constant state of stress and it's depressing her immune system," her communicator reports. Other pets dictate poetry to their owners telepathetically, we are told, and a cat at an animal refuge in New York state conceived and "communicated" the design for a new building. One communicator offers a course entitled, "What Animals Can Teach Us About Death."

THE HUMAN POLITICS OF ANIMAL CONSCIOUSNESS

The animal-rights and "deep ecology" literature is full of all sorts of similar talk about reconnecting "psychologically" and "spiritually" with animals to save the planet. But here one suspects the commitment is both emotional and political; these are the people out to knock man off his anthropocentric pedestal, after all, and what better way to do it than by setting up animals as our spiritual equals—or betters. This line of argument has even been adopted by some animal behavior scientists, including Sue Savage-Rumbaugh, an ape-language researcher who has advanced some of the more extravagant claims for humanlike mental abilities in apes. Recently she said that those who refuse to acknowledge the near-humanness of apes have "deluded" themselves in a desperate attempt to cling to "some small measure of safety." She continued: "But at the expense of gaining some comfort, we risk alienating ourselves pychologically from all other creatures on this planet."

The political agenda behind claims for animal consciousness seems at times to have become the driving force and raison d'être for such research. "If explorations of the minds of chimpanzees and other animals do nothing more than inform the debate about the ethics of animal use research, the work will have been well worth while," wrote a reviewer of a recent book on vervet monkeys. Jane Goodall, of chimpanzee fame, wrote in the foreword to a book by animal-rights philosopher Bernard Rollin that "it is the growing moral concern for animals and their welfare among the general public that is putting pressure on scientists to investigate animal consciousness and suffering."

Indeed, Savage-Rumbaugh does not hesitate to draw political conclusions from her research: Apes should be given "semi-human legal status," she says. She also insists that shifting the moral and legal boundary we have erected between man and other animals will help eradicate the attitudes that have "led us blindly to exploit the world of nature," destroy tropical rain forests, and mistreat animals.

Other researchers accuse Savage-Rumbaugh of overinterpreting the results of her language studies; where she sees her chimpanzees creating true sentences that display an understanding of syntax and semantics (the chimps push buttons labeled with symbols that stand for various nouns and verbs), others point out the extremely rote nature of the chimps' sentences, 96 percent of which are demands for food, toys, or tickling. As one of the chief critics, Herbert Terrace, has pointed out, interpreting a sequence of four button-pushes to be the equivalent of a syntactical sentence "please machine give M&Ms" makes no more sense than interpreting the sequence of buttons a person pushes while operating an ATM machine "as a sentence meaning, *please machine give cash.*" This is an issue we shall return to in chapter 6.

Savage-Rumbaugh counters that if anyone is politically motivated and irrational, it is those in the scientific orthodoxy who have an "emotional" fear of upsetting man's self-appointed place at the top of the animal heap and who accordingly have ruled out any inquiry into animals' interior mental experiences—their thoughts, feelings, and intentions. In making this charge she joins with a small but growing number of animal behaviorists who see the long-standing ban on anthropomorphism as a sort of conspiracy to disallow any evidence that might question man's special place in the world. "The concept of mind as we human beings experience it has for many people come to represent an unbreachable boundary between humans and nonhumans," she states, a boundary that "is being policed" by scientists. Others are more explicit in the conspiracy-theory charge; Donald Griffin, a Harvard researcher who is among the leaders of those decrying what he calls the "behavioristic taboo," says that animal behaviorists are "constrained by a guilty feeling that it is unscientific to study subjective feelings and conscious thoughts." He continues: "We have been so thoroughly brainwashed by the vehement rejection of suggestive evidence of animal thinking that it is considered foolhardy for students and aspiring scientists to let their thoughts stray into such forbidden territory, lest they be judged uncritical or even ostracized from the scientific community."

Imputing humanlike feelings and consciousness to animals is merely self-evident common sense that ordinary people have known all along, the conspiracy theorists argue; scientists who reject this view are just following the intellectual fashion of behaviorism, which, at least in caricature, posits that animals are nothing but stimulus-response machines incapable of even having mental states.

Most animal rightists trace the conspiracy back to Descartes, the great seventeenth-century champion of dualism, who, we are told, argued that animals truly were automatons, clockwork machines that lacked a mind altogether. But that view of clockwork, machinelike animals has been rejected by modern cognitive scientists, and indeed it is something of a caricature of even Descartes' views. Descartes' dualism was not so much a distinction between animals and man as between body and soul. The soul—possessed uniquely by man—was to Descartes a matter beyond scientific scrutiny, but the body—possessed by *both* man and animals—was fair game for science.

Even as they reject Cartesian dualism, though, modern cognitive scientists for the most part remain furious critics of anthropomorphism. Even the early twentieth-century behaviorists were not so much waging a philosophical war against the notion of continuity as they were waging a methodological war against the fuzzy thinking of "mentalism"—the uncritical late-nineteenth-century effort to impute conscious reasoning powers to animals. While mentalism perforce implies continuity, the opposite statement—that behaviorism implies dualism—does not hold at all. The behaviorists were not saying that animals do not have mental experiences, only that it is impossible for us to know. Given the repeated, and in retrospect often foolish mistakes made by researchers who enthusiastically mistook rote learning in animals as feats of conscious reasoning, this was a prudent course of action.

The behaviorists' fundamentally skeptical stance was well summarized by Edward L. Thorndike, one of the first true experimental psychologists who, in the early years of the twentieth century, noted that most books on animal behavior

> do not give us a psychology, but rather a *eulogy*, of animals. They have all been about animal *intelligence*, never about animal *stupidity*. . . . The history of books on animals' minds thus furnishes an illustration of the well-nigh universal tendency in human nature to find the marvelous wherever it can. We wonder that the stars are so big and so far apart, that the microbes are so small and so thick together, and for much the same reason wonder at the things animals do. Now imagine an astronomer tremendously eager to prove the stars as big as possible, or a bacteriologist whose great scientific desire is to demonstrate the microbes to be very, very little! Yet there has been a similar eagerness on the part of many recent writers on animal psychology to praise the abilities of animals. It cannot help leading to partiality in deductions from facts and more especially in the choice of facts for investigation. How can scientists who write like lawyers, defending animals against the charge of having no power of rationality, be at the same time impartial judges on the bench?

The only behaviorists who insist that the methodological decision to study only overt behavior actually reflects an underlying psychological reality—that is, that not only must we *study* animals as if they were stimulus-response machines, but they actually *are* stimulus-response machines—are the so-called radical behaviorists. Yet theirs is actually the ultimate continuity stance, for radical behaviorists argue that even human thoughts are nothing but conditioned responses—that is, that thinking actually *consists* of behavior, and that consciousness in truth does not exist. Or, as the psychologist Jeffrey Gray put it, "One is tempted to add: it is 'just a figment of our imagination.'" Gray also reported that he once asked a radical behaviorist what the difference would be between two awake individuals, one of them stone deaf, who are both sitting immobile in a room listening to a recording of a Mozart string quartet. The radical behaviorist answered: their subsequent verbal behavior.

One of the themes we shall return to is that between the annihilation of thought by the radical behaviorists and the elevation of thought to the level of human consciousness by nineteenth-century mentalists and modern "cognitive ethologists" like Griffin, lies a rich world of possibility. Modern cognitive science and evolutionary ecology are beginning to show that thinking in animals can be complex and wonderful in its variety, even as it differs profoundly from that of man.

I OPEN LATCHES, THEREFORE I AM

The historical roots of the modern distrust of anthropomorphism are worth a closer look, for the charge that it is all a "taboo" or irrational prejudice has tended to stick in recent years.

In the second half of the nineteenth century a number of writers, spurred on by Darwin's theory of evolution, began enthusiastically recounting tales of the power of reason displayed by animals—mainly clever dogs and other household pets. One of the greatest collectors of clever dog stories in support of Darwinian continuity was George Romanes, a friend and supporter of Darwin's. His book *Animal Intelligence* is crammed with anecdotes of dogs and cats that open latches and doors, that trick their owners with clever deceptions, and so on. But what distinguishes Romanes' writing is not so much the anecdotal nature of these illustrations as his enthusiasm for interpreting every feat of cleverness in terms of conscious reasoning—which he defined as an ability to perceive ratios or analogies, draw inferences, or predict probabilities. "We can only conclude," he wrote of latch-opening cats, "that the cats in such cases have a very definitive idea as to the mechanical properties of a door, they know that to make it open, even when it is unlatched, it requires to be *pushed* . . . she must reason, 'if a hand can do it, why not a paw?'"

Romanes' operating assumption was that "whenever we see a living organism apparently exerting an intentional choice, we may infer that it is a conscious choice." For example, he describes a fox caught by a farmer in a henhouse; the fox collapses on the floor and plays dead; the farmer chucks the seemingly lifeless body out the door; whereupon the fox gets up and runs away. Romanes concludes: "It seems to me that the probability rather inclines to the shamming dead having been due to an intelligent purpose."

Romanes acknowledged that to interpret an animal's thought processes this way required a heavy dose of inference from our own mental patterns, which we access through introspection. But Romanes invoked continuity to defend this process, which he termed an "ejective" method of inquiry, neither subjective nor objective, that permitted us to project our "own subjectivity" onto "the otherwise blank screen of another mind." He said a skeptic would be logically bound to deny evidence of mind in any organism, even in humans other than the skeptic himself.

Lloyd Morgan, another nineteenth-century pioneer of animal behavior study, agreed that introspection was the only source we have for "direct and immediate acquaintance" with psychological processes. But he also warned that applying our introspective knowledge to other animals was a "doubly inductive" process, and he was highly critical of wanton claims of reasoning in animals on the basis of feats that could be explained through simple learning alone, what Morgan called "sense-experience." He used his own anecdotes to show how easily we are misled, both by the very anecdotal nature of all of these stories and by our ready anthropomorphic projections onto the mental processes of the animals involved. He described his own Scotch terrier pup's struggles with carrying a cane through a gate. The dog had learned to carry the cane happily whenever he went for a walk, grasping it in the middle so it balanced comfortably. The first time they came to a gate, the dog dropped the cane and went on through. When sent back to fetch it, the dog seized the cane by its end and dragged it through. If the tale had stopped there, it would have been a perfect clever dog story: the animal had reasoned that the gate was too narrow for the cane to pass through if he held it by the middle, so he invented on the spot a novel strategy to cope with the situation.

But the tale didn't end there, for as Morgan was returning home with his dog along the same path, the animal did not drop the stick when they came to the gate, but rather repeatedly tried to smash through the gate holding the stick in the middle, each time striking the gate posts. Morgan then tried some more deliberate experiments, trying to teach a dog how to pull a crooked stick through a fence with narrow vertical rails. Morgan spent half an hour trying to show the dog how easy it was to pull the stick through, but each time the dog would try to yank it straight through and the crook would catch. Finally the dog seized the crook and as chance would have it broke it off. A passing man who saw only this end result stopped and observed to Morgan: "Clever dog that, sir; he knows where the hitch do lie."

As for the much-commented-on latch-opening abilities of animals, Morgan noted that a Scotch staghound that had learned to open a door into the yard performed the feat each time in precisely the same fashion: he would jump up on the door and scratch violently from the top downward over the entire area of the door, until at some point or other he finally struck the latch and the door opened. The dog had obviously done this once and it had worked, and then had stuck with a proven formula, a formula that betrayed not the slightest grasp of the underlying principle of the latch.

Morgan was careful to emphasize that he was not adopting what he called "the false position of dogmatic denial of rational powers to animals." But he noted that again and again the stronger claims for reasoning by animals, made by projection of human mental experience, did not stand up to scrutiny. Animals did remarkably stupid things in situations completely analogous to situations where they had exhibited apparent insight or reasoning; they demonstrably learned some of their clever feats by pure accident; and anecdotes could never tell you what previous learning experiences an animal might have had before performing its seemingly clever feat. It was not dogma but hard experience that led him to formulate "Morgan's canon," his "basal principle" that "in no case may we interpret an action as the outcome of the exercise of a higher psychical faculty, if it can be interpreted as the outcome of the exercise of one which stands lower in the psychological scale."

Edward L. Thorndike extended Morgan's criticisms. His chief point was that it was not merely unnecessary to invoke reasoning by analogy or other conscious thought processes to explain animal behavior; it could often be shown on further investigation that such explanations were actually wrong, however tempting and superficially convincing they at first seemed to be. Thorndike's method was to study how animals learned completely novel tasks under controlled conditions. In his most famous experiments, he placed cats in "puzzle boxes"; to escape, the animals had to press a lever or pull a string or sometimes perform a series of such actions. On their first try, all of his cats showed "customary instinctive clawings and squeezing and biting." In one test where cats had to press a thumb latch and push against a door to escape, eight cats did manage in the course of their (clearly random) struggles to push down the thumb piece; only six ever managed to push the thumb piece and press against the door at the same time; and of those only three, and only after repeated trials, managed to associate their actions with escape. "The great support of those who do claim for animals the ability to infer," Thorndike concluded, "has been their wonderful performances which resemble our own. These could not, they claim, have happened by accident. No animal could learn to open a latched gate by accident. The whole substance of the argument vanishes if, as a matter of fact, animals do learn those things by accident. *They certainly do.*"

CLEVER HANS, CHIEF SPOILSPORT

There was no more dramatic illustration of the point Thorndike was making than Clever Hans, a horse who was making headlines in Berlin at almost the same time Thorndike was writing those words. Clever Hans has since become a staple cautionary tale in animal behavior research. Hans had gained his fame by solving mathematical problems—and not just simple arithmetic, but also puzzles such as adding 25 to 15 or finding the factors of 28. Hans could also tell time, identify musical scores, and answer questions about European politics. He would count out the answers to the math problems with his foot and answer "yes" or "no" to other sorts of problems by nodding or shaking his head. His owner, an elderly schoolmaster named Wilhelm von Osten, had patiently taught Hans his lessons, rewarding him with a sugar cube when he got the right answer. Von Osten was no perpetrator of hoaxes but an honest man who firmly believed in his horse's mastery of his subjects.

Von Osten was hardly the only believer. As one psychologist noted at the time, many zoologists saw in Hans's abilities confirmation of the essential similarity between human and animal minds, a doctrine that "has been coming more and more into favor since the time of Darwin."

At last, an investigation by the Prussian Academy of Sciences got to the bottom of Hans's remarkable talents. Remarkable they were, but not in the way anyone had believed. The key finding was that Hans could not answer questions correctly when no one in the room with him knew the correct answer—for example, when two people separately whispered numbers to Hans for him to add together, but did not tell each other their numbers until after Hans had had his chance to answer. What was happening was this: Unconsciously, Hans's questioners were cueing the horse, for example by subtly bobbing their heads in anticipation of the correct answer, or tensing up as Hans counted and then subtly relaxing when he got to the right number. People who knew the right answer ahead of time were naturally anxious to see if Hans would get it right, and were betraying some sign of acknowledgment, unconscious even to themselves but not to Hans, when he did.

Horses, as social, herd-dwelling animals adapted to an open environment, have a remarkable evolved ability to pick up on subtle visual cues from their fellows. Hans was undoubtedly "clever" in that regard. He was able to form a myriad of very subtly linked associations through learning. He had discovered that if he stopped pawing just when the appropriate cue appeared, he would get a sugar cube. But he did not in fact "know" a thing about square roots, the kings of Spain, or Beethoven.

The lesson was—or should have been—simple: We are easily fooled by animals' ability to learn from inadvertent cues. We are especially easily fooled when that learning takes a form that, on the surface, appears identical to things that people do.

Neo-mentalists such as Savage-Rumbaugh indignantly complain that the story of Clever Hans has become part of the conspiracy to rob animals of their due. James Gould, a protégé of Donald Griffin's, has said that "once the Clever Hans story began circulating, any suggestion that animals had any native intelligence was in trouble. Behaviorists began using the incident as proof that everything nonhuman creatures do is simply the result of instincts programmed into them from birth." But this is an odd complaint. No scientists ever suggested that Clever Hans hadn't *learned*, or that he was merely following preprogrammed instincts. He certainly had learned. He just hadn't learned what the quick-to-anthropomorphize humans *thought* he had learned.

In the Clever Hans episode, anthropomorphism wasn't so much the original sin as the cover-up for the sin. The original sin was an experimental method that permitted inadvertent cueing; what anthropomorphism did was to provide a superficially convincing explanation that could cover a multitude of such sins. The reason that most behavioral researchers reject anthropomorphism is precisely because it offers a pat explanation that lets researchers off the hook from probing deeper for alternative explanations or confounding variables. Imputing reason and understanding seemed to explain Clever Hans's behavior perfectly; why look further?

Inadvertent cueing is just one of many confounding factors that further investigation *has* shown up. A related factor that bedevils many experiments (including, as we shall see, many experiments specifically designed to probe cognitive processes) is imprecise experimental design. The experiment mentioned at the beginning of this chapter, in which capuchin monkeys were tested for their ability to categorize "person" and "nonperson," is a good case in point. The initial results seemed very promising; the monkeys were able to correctly categorize

75 percent of the novel slides they were shown. But, ever mindful of Clever Hans, the researchers realized from the start that there was simply no guarantee that an animal's seemingly correct performance was controlled by the same categories conceived of and perceived by their human experimenters. The experimenters tried to control for as many inadvertent cues as they could think of. Maybe the monkeys weren't responding to the image in the picture at all, but to its overall brightness. Maybe there were objects in the background of the photos that the monkeys had picked out to form their "categories." It was precisely those doubts that led the experimenters to press on even after obtaining their strongly positive results. And what they found was a classic Clever Hans phenomenon: a subsequent analysis of the mistakes the monkeys made revealed something very funny. A significant proportion of the slides incorrectly categorized by the monkeys as a "person" had a patch of red somewhere in the image. And the nonperson slides most likely to be misidentified as a person had that patch of red as a feature of an animal or a flower. In other words, the monkeys seemed not to be using the experimenters' categories at all, or at the very least they were categorizing the images on a different set of criteria from what the humans who designed the test imagined.

Another confounding variable that anthropomorphic interpretation can disguise is previous learning or experience. This proved to be the case with Wolfgang Köhler's famous experiments with chimpanzees in the 1910s. Köhler noted that in Thorndike's puzzle box experiments, the animals were forced by the design of the escape mechanisms to resort to trial and error, as the mechanism itself was hidden. Köhler decided to present animals with problems where the solution was in plain view, to see whether they were capable of working out the solution by insight rather than chance. In one test, for example, food was placed outside of a barred cage, out of reach of the chimpanzee within. The chimpanzee was supplied with a stick, however, that could be used as a rake to pull the food within reach. In a more difficult variation, the food could be raked in only if two sticks, which could be fitted together end to end, were used. Köhler concluded that some of his chimpanzees did indeed use insight to solve this problem. The difference between

Thorndike's and Köhler's animals led Bertrand Russell to wryly comment that "All of the animals that have been carefully observed . . . have all displayed the national characteristics of the observer. Animals studied by Americans rush about frantically, with an incredible display of hustle and pep, and at last achieve the desired result by chance. Animals observed by Germans sit still and think, and at last evolve the solution out of their inner consciousness."

But it has since become apparent that Köhler's Germanic chimps did *not* evolve the solution out of their inner consciousness. Chimpanzees that have never had the chance to play with sticks fail to use the sticks as rakes in such experiments. And chimpanzees supplied with sticks even when there was no problem to solve immediately began playing with them; such "irrelevant" behavior is extremely common in apes. (In one experiment, forty-eight chimpanzees were given sticks that could be fitted together; within an hour thirty-two of them had done so.) It was only by first having had such a chance to learn, by trial and error, what a stick can do that the chimpanzees were able to use them to solve problems. Once again, the readiness to accept insight as an explanation for animal behavior—a very human (and not just Germanic) interpretation of events, as we know we do in fact solve problems that way—led to experimental blindness.

Invoking chance and coincidence to explain away seemingly impressive cognitive feats by animals may seem like a cop-out criticism, a counterargument you use when you can't think of anything else. But it is absolutely relevant in the case of anecdotes—especially anecdotes plucked out from perhaps thousands of hours of observations precisely because they seemed remarkable examples of humanlike thought or strategy or problem solving. It is not just a pedantic criticism to note that lots of things that look clever really *are* nothing more than chance. If you pick a winning horse at the track, you can have great fun "analyzing" for your friends the brilliant calculations you made and the expert knowledge you brought to bear on your choice, when in fact you made your selection based on the fact that the horse had the same name as your old girlfriend. And lots of people place winning bets every day.

In one much repeated anecdote, Washoe, the first chimpanzee to be "taught" sign language, was reported by her trainers to sign *water* plus *bird* when she saw a swan. It was a novel combination and seemed to show a creative insight. Maybe it did, but given the number of inane and meaningless (or excruciatingly repetitive) signs Washoe made, it is hardly surprising that one or two novel combinations should appear to make sense—especially given the certainty that any such coincidences that did occur would be eagerly seized on and reported, while the inanities would not. As others have pointed out, too, there was both water and a bird present in Washoe's environment when she signed *water bird*, so there may be an even simpler explanation.

Clever Hans illustrated vividly how the very act of training an animal with the aim of demonstrating an advanced mental ability can queer the pitch. A researcher out to prove his case is terribly susceptible to betraying his desired results to his animals, who, as Clever Hans showed, are remarkably adept at producing more-than-reasonable facsimiles of an ability his trainer wishes to find. Treat your dog like a human, and he tries to be like one, precisely because you reward him with attention for his human-like behavior. If you ooh and ahh when your dog sits with rapt attention in front of the television whenever Pavarotti is on, he will easily make the association between his behavior and the reward. Some dogs are wonderfully quick to learn to throw up regularly if they have been rewarded by being fussed over when they are sick. This happens often enough that veterinarians refer to it as the "sick pet syndrome." Researchers who push mentalist explanations are understandably on edge about always having to defend themselves against this criticism, but it's not as if it is a groundless one. Those who ignore Clever Hans are doomed to have their monkeys make monkeys of them.

I

WHO IS THE SMARTEST
OF THEM ALL?

THE QUESTION ALMOST EVERYONE ASKS WHEN THE matter of animal minds comes up is: How smart are they? It is a question that occupied early researchers, too, at the dawn of experimental comparative psychology a century ago. Because intelligence seemed to be a quantifiable, testable parameter, studying an animal's performance on problem-solving tasks seemed the most promising avenue to exploring their mental processes.

Ranking species according to their relative intelligence fit in well, too, with many popular—though often terribly distorted—conceptions about Darwin's theory that were current in the late nineteenth century, and which have not yet altogether vanished. One of the most enduring misperceptions about evolution is that life represents a sort of chain of progress from inferior to superior forms. A particularly muddled version of this embodies the additional notion that progress up the chain of evolutionary advancement corresponds to the steps that the most advanced forms of life (i.e., humans) follow in the course of development from infancy to adulthood. This theory that "ontogeny recapitulates phylogeny" has no credence whatsoever in modern biology. Yet it still pops up regularly, and certainly most people tend to think about evolution in these terms, as a stepladder along which species can be ranked as higher or lower.

1

George Romanes, the great nineteenth-century collector of animal anecdotes, constructed an elaborate chart of the relative rankings of animals' stages of "mental development" that is a paradigm of this sort of thinking. To make the picture of a stepladder of mental ranks even neater, he equated each step with a corresponding age of mental development in humans. Thus a portion of his ranking chart looked like this, with the rank number in the first column and the corresponding stage of human development in the last:

28. Indefinite morality.	Anthropoid Apes and Dog.	15 months.
27. Use of tools.	Monkeys, Cat, and Elephant.	12 months.
26. Understanding of mechanisms.	Carnivores, Rodents, and Ruminants.	10 months.
25. Recognition of pictures, Understanding of words, Dreaming.	Birds.	8 months.
24. Communication of ideas.	Hymenoptera. [Bees, Ants]	5 months.
23. Recognition of persons.	Reptiles and Cephalopods.	15 months.
22. Reason.	Higher Crustaceans.	14 months.
21. Association by similarity.	Fish.	14 weeks.
18. Primary instincts.	Larvae of Insects.	3 weeks.
17. Memory.	Echinodermata [Starfish, etc.]	1 weeks.
7. Non-nervous adjustments.	Unicellular organisms.	Embryo.
3. Protoplasmic movements.	Protoplasmic organisms.	Ovum and Spermatozoa.

ANIMAL IQ

There are two basic flaws with this approach. One is the fundamental notion that some species are more highly "evolved" than others. We might naturally think of monkeys occupying a position higher on the phylogenetic scale than cats, and cats higher than rats. Yet the fact is that primates, carnivores, and rodents all diverged from a common ancestor at the same time. They are all equally "evolved." The branching tree of evolution has not just one culmination, but millions of culminations—represented in every living species on earth today. Each is a brilliant success at what it does. The idea that fish, now stuck at level 21, are trying with all their might to ascend to level 22 is, from an evolutionary point of view, nonsense. Fish are adapted by virtue of millions of years of evolution to their particular, special evolutionary niche. They have had just as long to evolve as we have. It is not as if they are mere instances of incomplete evolution, the culmination of which is man (or Nordic man, perhaps).

The other problem is the implicit assumption that intelligence is something that can be measured on a linear scale—and a scale where humans equal 100. Of course, by explicitly defining his stages of mental evolution in the animal kingdom in terms of the stages of mental development in human infants, Romanes was bound to end up with a purely anthropocentric definition for his rankings of intelligence. The "Properties of Intellectual Development" he chose to list (morality, use of tools, recognition of persons, etc.) all have a distinctly self-centered air about them. How would one fit the navigational abilities of pigeons or the web weaving of spiders or the nest building of bowerbirds or the food caching of nutcrackers into Romanes's scheme? One wouldn't.

Some modern attempts to define intelligence in universal terms do not fare much better. Most common definitions of intelligence emphasize flexibility, creativity, and recognition of underlying patterns and overarching concepts. Some researchers, such as Steven Pinker of M.I.T., define intelligence in a more restrictive way that would seem to rule animals out of the running altogether: Pinker says intelligence is an ability to figure out how things work in order to overcome obstacles.

But animals, which as we shall see do not show any notable ability to figure out how things truly work, nonetheless show great facility at accomplishing things by acting on information they receive from the environment. They make decisions that are flexible and often appropriate. Unconsciously operating algorithms in the animal mind (ours included) produce what we would not hesitate to call intelligence were we to see a robot do it. Coordinating the movement of four legs over uneven ground while avoiding obstacles is a sophisticated computational task. We do not normally think of such automatic tasks as part of "intelligence," but why not? From a purely computational viewpoint, the sort of unconscious thought that permits an animal to recognize a predator and take evasive action could certainly involve far more brainpower than distinguishing a group of three items from a group of four items. Is the former just dumb reflex and only the latter intelligence?

Another huge problem in attempting an honest, "zero-based" assessment of animal intelligence is the bias and assumptions built into the many tests that have been devised to measure it. Intelligence tests for humans have long been criticized for being culturally biased. Many of the early IQ tests used in the United States in particular suffered from this fault; they claimed to measure "native intelligence," but included questions that really measured nothing so much as familiarity with the middle-class American culture of the middle-class American psychologists who devised the tests. For example, tests given to Army recruits in World War I featured such questions as, "Washington is to Adams as first is to . . ." Other questions required the examinees to draw in missing parts on a series of pictures—a stamp on an addressed envelope, a net on a tennis court, a filament in an electric light bulb, a horn on a wind-up phonograph, a trigger on a pistol, strings on a violin. Not surprisingly, many recent immigrants to America did not score very highly on the exam and were rated as "morons" or "feeble-minded." Probably not very many of them played tennis, either.

But differences between human cultures pale in comparison to differences between animal species. Animals differ in temperament, perceptual abilities, motivation, social behavior, all of which affect their performance on tests we might devise for them.

WHO'S SMARTER: THE SHEEPDOG OR THE SHEEP?

Sheep have a reputation for being dumb. Border collies have a reputation for being smart. But both of these impressions may say more about our underlying prejudice than their underlying intelligence.

Much of what impresses us about dogs, after all, is their obedience to us. To put it in a slightly cynical fashion, we say a dog or a horse is smart when it does what we want it to. But many disobedient dogs or horses—the ones we curse as stupid—are actually every bit as clever as the "smart" ones. As quickly as the smart ones learn obedience the stupid ones learn evasions. We tend to be impressed by a dog that learns to run around behind a flock of sheep and instantly lie down on command. We are totally unimpressed by a dog that has gone to six weeks of obedience school only to begin selectively ignoring the command "here" whenever he is busy sniffing and digging and urinating in the neighbor's yard. But in both cases the underlying "intelligence" is arguably the same. In fact, from a formal learning point of view, the latter could even be seen as a superior feat—for the dog has learned a conditionality task here: Come when called, except when you're far away and the immediate rewards of ignoring the command are greater than obeying.

Disobedient horses have likewise often mastered a sophisticated learned association—just not the one we're trying to teach them. A disruptive horse whose antics scare its rider and cause him to end the lesson and take the horse back to the barn has cleverly learned to associate wild behavior on its part with the immediate reward of not having to work anymore. From our point of view, the horse is unteachable; in fact it is a very good learner.

The difference between the "smart dogs" and "dumb dogs" is thus largely a matter of differing temperaments—and an artifact of our self-serving definition of intelligence in dogs. Dogs that exhibit dominant behavior are not teachable because they resist our authority. Horses that are too timid or fearful are not teachable because they tend to react emotionally and seek to escape altogether from situations when they are corrected. If most dogs and horses are teachable, that is largely because they are highly social animals, attuned to vocal and physical signals of dominance and submission; they both have a natural instinct to defer to the will of a pack or herd leader. Few of the things we ask them to do go completely or even partially against their natural propensities. Dogs in general can be house-trained because they are den-dwelling animals that have a strong instinct to keep their own nest clean. They can be taught to "shake hands" because raising a paw in that fashion is a natural submissive gesture. Border collies can be taught to herd sheep because they have been selected for generations to emphasize the part of the natural hunting instinct of wolves that involves circling and cutting off fleeing prey. People often remark how smart a Border collie is not to attack and eat the sheep—how this goes against its natural instinct. It does nothing of the kind. Many dogs show no interest in chasing prey at all, after all; livestock guard dogs such as Great Pyrenees and Maremmas have been bred to relate to sheep more as littermates than as prey and to behave aggressively toward intruders. Cleverness doesn't enter into it.

As for the sheep we dismiss as stupid, they are quite adept at learning and recognizing individuals by their faces; they quickly and perceptively catch on to new feeding schedules; they are good at finding holes in fences. And while people do not generally think of sheep as animals that can be taught anything, in fact farm flocks can easily be trained in a few lessons (with suitable food rewards) to come when called, and show sheep are routinely trained to walk on a lead and to stand still while a judge inspects them.

Although, like horses and dogs, they are social animals with a dominance hierarchy, sheep are very shy and fearful creatures—a clearly adaptive trait for a small, prey animal—and they lack the expressive repertoire of vocal and physical signals that these other species have, relying more on direct physical contact to establish and reinforce the hierarchy. All of which means that it is simply harder to teach them to learn the associations we might want to teach them. However we are setting up a false dichotomy here, for the fact is that there is not very much we *want* to teach sheep to do. Their temperament and size just do not lend them to the roles filled by horses and dogs (a watch sheep?) so we just do not bother training them. If a person raised a lamb in the house and could think up things to teach it, there is no particular reason to doubt that it would learn them.

It is interesting that we are ready to make allowances for cats that we won't make for sheep: Most cat owners vouch for the intelligence of their pets while conceding their untrainability. The evolutionary explanation works well here, too: Cats are not group-dwelling animals, and have had no need in their evolutionary history to develop the complex system of dominance and submission that, together with subtle threats and appeasement gestures, serves so well to keep the peace in a large group. The role we can take on as pack leader or herd leader in persuading a dog or a horse to give in to our will without a fight does not exist in the cat's ecological niche.

In fables and legends, humans have tended to rate as intelligent those animals that simply have a keen visual sense (the wise owl) and dexterous hands (foxes, monkeys). Even most of the "scientific" rankings of intelligence that have been attempted largely appear to be nothing more than a rank order of visual acuity. We are visual animals ourselves, and animals that can see what we see seem smart; those that rely more on their sense of smell or hearing are the ones we tend to think of as dim bulbs. Horses tend to shy a lot because the construction of their eyes is optimized for a near 360-degree field of view, useful for spotting danger, but the price the horse pays for that is relatively poor acuity and some out-of-focus spots that can cause objects within the field of view to suddenly sail into sharp focus. "You stupid horse," we say when the animal suddenly jumps at the sight of a mailbox that it seemed to have been looking at for some time; "you horse lacking the acuity typical of the human visual system" would be a more just epithet.

Our bias in favor of animals that can see and do things is a very fundamental point we have to face up to in trying to assess relative intelligence. "If a goldfish was as intelligent as a chimpanzee, how would it show it?" asks psychologist Euan Macphail of York University in England. "A goldfish doesn't have any limbs, it doesn't have a very good visual system. When we set it a learning task, though, it learns perfectly efficiently." An animal's input and output devices, in other words, inevitably affect our impression of the power of the central processor. If there is such a thing as general intelligence or (say) general decision-making ability, it still can only work on the inputs it receives and can only show itself through the outputs that reach the world at large. Macphail's point is that the reason monkeys and apes use tools may be nothing more complex than that they have hands. You might have the latest computer that can run an action game with elaborate 3-D full-motion color video, but if you hook it up to a monochrome monitor and a paper-tape reader, you would never know it.

The long obsession with brain size as an indicator of animal—and human—intelligence reflects the belief that intelligence is some general-purpose "stuff" that an individual organism is endowed with more or less of. Correlation between intelligence and brain size has been soundly rebuffed in humans, and ought pretty obviously to be equally insignificant a factor in animals; both from individual human to individual human and species to species, the factor that has by far the strongest correlation to brain size is body size. Nobody would seriously argue that tall men are smarter than short men, or that elephants are several times smarter than dogs because their brains are several times larger. One simple reason brain size grows with body size is that a larger animal has more sensory nerve fibers coming into the brain from all over its larger body, and more nerve fibers leaving the brain to control the muscles. Another point to recognize is that as brains get larger, they do not even necessarily increase their number of nerve cells in proportion. As the distance between nerve cells grows, so does the thickness of the "wires" that connect them. If they didn't, the nerve signals would attenuate too much by the time they had traversed this greater distance. So more of the space in bigger brains is taken up with the "wiring," and to a first approximation, the number of nerve cells is actually independent of total brain size. Bigger brains are simply less dense.

Of course that is not the entire story; differences in brain structure and organization do begin to appear when you start comparing reptiles and birds or monkeys and cats, or people and chimpanzees. Horses, carnivores, monkeys, apes, and man all have brains that fall significantly above the body-size/brain-size curve. (Humans have brains three times as large as a proportion of body weight as do other primates.) The major difference, though, is the increased size of the neocortex in monkeys, apes, and man. This sheet of nerve tissue, which wraps around the brain of mammals, is the so-called gray matter, and plays the lead role in processing information from the sense organs and in controlling the movement of the limbs. In hedgehogs, the neocortex accounts for less than 30 percent of the brain's total volume; in monkeys it is about 70 percent, in chimpanzees 75 percent, and in humans 80 percent. If the cortex were just a flat piece of newspaper wrapping a grapefruit, its total volume would increase more slowly than the grapefruit as a whole with increasing grapefruit size. Instead, the cortex gets increasingly wrinkled so its total surface area increases dramatically. Flatten out the cortex of a human, William Calvin notes, and it would cover four sheets of typing paper. A chimpanzee's would cover one sheet, a monkey's would cover a postcard, a rat's a postage stamp.

Neurological studies have found that different areas of the neocortex are devoted generally to different functions and there is a dramatic correlation between the size of each area devoted to each sense organ and the importance of that sense to the animal. Monkeys, diurnal animals that have a high visual acuity—necessary for finding food and for moving through the trees without bumping into things or missing one's hold on a branch—have a large visual area of the neocortex. Cats, nocturnal animals that rely heavily on hearing to find prey, have a correspondingly large auditory area. The touch and motor-control regions of the neocortex in humans have very large areas devoted to touch and motor control of the hands and fingers; New World monkeys have correspondingly large areas devoted to the end of the tail, which they use for grasping. Although birds lack a neocortex, their brains possess corresponding structures that appear to fill the same roles that the neocortex does in mammals.

Other parts of the neocortex appear to be involved in learning; even these "association areas" of the mammalian cortex appear to have their specialties. If one part of a monkey's neocortex is removed, it has difficulty learning visual tasks, such as which of two containers is the one with food; other parts affect its ability to learn to discriminate by sound or by touch.

ARE INPUT AND OUTPUT PART OF INTELLIGENCE?

Reptiles and amphibians clearly can learn (even worms can!), and attempts to show that a bigger neocortex or a bigger association area is correlated with greater learning ability or greater general intelligence have proved to be anything but straightforward. There are two closely related problems we encounter here—one methodological, the other philosophical. The philosophical problem is that, as we have seen, we lack a good definition of intelligence. The traditional views on the subject, while disagreeing on many things, generally take as a common starting point that intelligence is what the brain does when it is stripped to the core. Control for the fact that animals have different sensing and manipulative abilities, and beneath it all is that general-purpose calculating and learning machine where intelligence resides. It seems natural that some animals have bigger and faster machines than others. And it also has been assumed that the basic workings of these machines all take the same form. If intelligence is a matter of *general* intelligence, then general problem solving and learning ability ought to be equally applicable to all problems. Thus comparative psychologists have traditionally used abstract learning problems such as matching samples or learning lists as a way to probe those underlying computational abilities, without regard to the particular body packages they come in.

More recently, though, artificial-intelligence researchers have increasingly come around to the view that a crucial part of intelligence is precisely those "peripherals" that allow the brain to operate in the real world. The point is that a superfast computer with gigabytes of memory really *is* dumb if it can only receive inputs from paper tape and output them on a black and white text-only screen—its intelligence is more like that of an idiot savant than a normal human, great at solving abstract problems (like chess) but rotten at having the more general awareness of the world, and ability to relate to real problems in the real world, that (so far, anyway) differentiate a robot from a person. The ability to do things in the real world depends on the ability to receive accurate and complete data and to be able to respond in an effective way.

This view leads to a conclusion that corresponds well with "common sense" judgments we make about the relative intelligence of animals. The fact that an ape has hands and can control them really does, after all, permit it to tackle problems that a goldfish cannot. But the reason a chimpanzee is good at using tools may have far less to do with any innate superiority in its core, information-processing ability than with the fact it has hands and the machinery to run them. "We don't have any reason to suppose that if dogs had hands they wouldn't use them," says Evan Macphail.

Part of what we take for intelligence, in other words, rests in the sense and motor organs and in the wiring that controls them. The neocortex is more than just a passive mechanical linkage between the sense organs and the real brains of the operation in the central processor. In this view, intelligence is the sum total of the processor plus the peripherals. If by intelligence we mean the ability to formulate effective solutions to variable environmental challenges, then this is clearly a correct view. By way of analogy, consider the challenge a foxhound pack faces. A fox leaves a meandering and highly variable scent trail. The pack spreads out across the territory, casting for a scent; when one hound finds the scent it barks, and the other hounds respond by racing to that spot and focusing their efforts in that area. This process is continually repeated to follow the fox's trail. The pack together is far more effective than a single hound would be in solving this problem, precisely because of the addition of many extra noses and a communication system that directs where those noses should be brought to bear. If you added only extra foxhound brains, you would not noticeably increase the "intelligence" of this distributed system. It is by hooking up extra peripheral devices that the network gets smarter.

The methodological problem is that whatever stance we take on what intelligence is, it is extremely hard in practice to separate out the processor from the peripherals experimentally. Whether you call it perception or whether you call it part of intelligence per se, the peripheral devices play such a large part in the performance of intelligent acts by animals that it is difficult to be sure we are comparing like with like when we try to measure the processor part of the equation. Giving a blind person a written IQ test is obviously not a very meaningful evaluation of his mental abilities.

Yet that is exactly what many cross-species intelligence tests have done. Monkeys, for example, were found not only to learn visual discrimination tasks but to improve over a series of such tasks—they formed a learning set, a general concept of the problem that betokened a higher cognitive process than a simple stimulus-response association. Rats given the same tasks showed difficulty in mastering the problems and no ability to form a learning set. The obvious conclusion was that monkeys are smarter than rats, a conclusion that was comfortably accepted, as it fit well with our preexisting prejudices about the distribution of general intelligence in nature. But when the rat experiments were repeated, only this time the rats were given the task of discriminating different smells, they learned quickly and showed rapid improvement on subsequent problems, just as the monkeys did.

The problem of motivation is another major confounding variable. Sometimes we may think we are testing an animal's brain when we are only testing its stomach. For example, in a series of studies goldfish never learned to improve their performance when challenged with "reversal" tasks. These are experiments in which an animal is trained to pick one of two alternative stimuli (a black panel versus a white panel, say) in order to obtain a food reward; the correct answer is then switched and the subject has to relearn which one to pick. Rats quickly learned to switch their response when the previously rewarded answer no longer worked. Fish didn't. This certainly fit comfortably with everyone's sense that fish are dumber than rats. But when the experiment was repeated with a different food reward (a paste squirted into the tank right where the fish made its correct choice, as opposed to pellets dropped into the back of the tank), lo and behold the goldfish suddenly did start improving on reversal tasks. Other seemingly fundamental learning differences between fish and rodents likewise vanished when the experiments were redesigned to take into account differences in motivation.

Equalizing motivation is an almost insoluble problem for designers of experiments. Are three goldfish pellets the equivalent of one banana or fifteen bird seeds? How could we even know? We would somehow have to enter into the internal being of different animals to know for sure, and if we could do that we would not need to be devising roundabout experiments to probe their mental processes in the first place.

When we do control for all of the confounding variables that we possibly can, the striking thing about the "pure" cognitive differences that remain is how the similarities in performance between different animals given similar problems vastly outweigh the differences. To be sure, there seems to be little doubt that chimpanzees can learn new associations with a single reinforced trial, and that that is genuinely faster than other mammals or pigeons do it. Monkeys and apes also learn lists faster than pigeons do. Apes and monkeys seem to have a faster and more accurate grasp of numerosity judgments than birds do. The ability to manipulate spatial information appears to be greater in apes than in monkeys.

But again and again experiments have shown that many abilities thought the sole province of "higher" primates can be taught, with patience, to pigeons or other animals. Supposedly superior rhesus monkeys did better than the less advanced cebus monkeys in a visual learning-set problem using colored objects. Then it turned out that the cebus monkeys did better than the rhesus monkeys when gray objects were used. Rats were believed to have superior abilities to pigeons in remembering locations in a radial maze. But after relatively small changes in the procedure and the apparatus, pigeons did just as well.

If such experiments had shown, say, that monkeys can learn lists of forty-five items but pigeons can only learn two, we would probably be convinced that there are some absolute differences in mental machinery between the two species. But the absolute differences are far narrower. Pigeons appear to differ from baboons and people in the way they go about solving problems that involve matching up two images that have been rotated one from the other, but they still get the right answers. They essentially do just as well as monkeys in categorizing slides of birds or fish or other things. Euan Macphail's review of the literature led him to conclude that when it comes to the things that can be honestly called general intelligence, no convincing differences, either qualitative or quantitative, have yet been demonstrated between vertebrate species. While few cognitive researchers would go quite so far—and indeed we will encounter a number of examples of differences in mental abilities between species that are hard to explain as anything but a fundamental difference in cognitive function—it is striking how small those differences are, far smaller than "common sense" generally has it. Macphail has suggested that the "no-difference" stance should be taken as a "null hypothesis" in all studies of comparative intelligence; that is, it is an alternative that always has to be considered and ought to be assumed to be the case unless proven otherwise.

ECOLOGISTS VERSUS GENERALISTS

One interpretation of these findings is that intelligence fundamentally consists of specialized intelligence, and that there may not really be much of anything left in the brain even worthy of the name general intelligence. Most of the genuine differences that appear in what different species can do reflect adaptations to the special ecological requirements of each. Spiders spin webs; many birds sing their songs even if they have never heard a model; cowbirds hunt out other birds' nests to lay their eggs in; male moose knock heads with other male moose. None of these are trivial cognitive tasks; all involve decision making, or perceptual processing, or both. All reflect highly specialized adaptations to a particular environment and way of life. The very fact that so much of the brain is devoted to the processing and control of the particular sense and motor organs that are important to an animal's way of life speaks to this point.

Learning itself seems to be highly specialized, too. Animals that can learn one sort of task often cannot learn a logically identical task. The ability to learn seems less a result of general cognitive processes than specialized channels attuned to an animal's basic hard-wired behaviors. The classic demonstration of this point came in the course of Edward Thorndike's pioneering experiments in which cats had to learn to escape from boxes by pushing levers or pulling strings. In one of his variations on this theme Thorndike placed his cats in "Box Z." This consisted of a box with no levers or gizmos at all, only a door that could be opened by the experimenter. He would do so whenever the cat in the box licked or scratched itself. The cats had considerable difficulty learning this association— far more than they had in learning to pull strings or press levers. Likewise it is essentially impossible to teach dogs to yawn for a food reward. Dogs can be taught to go to the left-hand box or the right-hand box to retrieve a reward depending on whether a speaker above or below the dog sounds a tone, but dogs find it extremely hard to learn whether to go or stay according to which speaker sounds a tone. Pigeons readily learn to physically flee a shock but find it extremely difficult to learn to peck a key to avoid a shock—even though they readily learn to peck keys to obtain food. (In fact, pigeons quickly learn to peck a lighted key even *without* a food reward.) Rats can learn to avoid a food that causes nausea hours later but would never make an association between a key press and a punishment so long delayed. The point is that many learned responses are "prepared" or "contraprepared" by the way an animal's brain is prewired—prewired precisely by species-specific ecological adaptations. Pecking at things is part of the basic feeding pattern of pigeons. Flight from a noxious stimulus is a basic part of most species' defensive repertoire. Animals with paws manipulate things in their environment. Learning appears less a general cognitive process that resides in some central processor than one channeled through—and inseparable from—the "peripheral" processing apparatus of the brain.

Macphail's notion that all animals possess an equal measure of "general intelligence" might at first blush seem to be the opposite pole from the ecological stance, which supposes that all intelligence reflects specialized adaptations to a unique environment. But the positions are far closer together than they might appear, and their apparent difference really goes back to the matter of how we define intelligence. Macphail's point is that if you strip away all perceptual and manipulative differences between species, what is left—which is what he defines as real intelligence—is pretty much the same across the board. The ecological stance in effect is saying that real intelligence predominantly consists precisely of all the stuff Macphail stripped away—the species-specific wiring of the sense and motor organs to the brain—thus the unique differences from one species to another. Some in the ecological camp have argued that even the fundamental learning processes differ from species to species and specialized task to specialized task; that there is one mechanism for learning about spatial locations of food sources and another mechanism for learning social hierarchies and so on. But the fact remains that when arbitrary learning tasks are devised in a way that is careful to avoid sensory biases, even very different animals often learn them the same way and with equal speed.

Either way, we are left with a very interesting conclusion: That all species (all birds and mammals anyway) are pretty much equally intelligent. Which species is the smartest? They all are. It is simply meaningless to ask which of the following reflects the greater intelligence: the calculations a pigeon's brain performs to find home or the calculations a wren's brain performs to determine the distance of a rival from the sound of its song or the calculations a chimpanzee's brain performs to assess which food cache it has visited in the past is the closest. Specialization means *by definition* that different abilities are simply not comparable. It is the classic "apples and oranges" situation; it is literally the same as asking, Which is better, a wing or an arm? Fur or hair? Two legs or four legs? Night vision or daylight vision? A short tail or a long tail? Nests or webs? Paws or hooves? Lungs or gills? The answer in every case is, of course, that it depends—what do you want to use them for?

And Macphail's additional point is that to the extent different abilities are comparable, they appear to rely largely—or possibly wholly—on a common mechanism. The differences in intelligence between animals is a matter of quality, not quantity.

WHAT ABOUT US?

Are humans able to acquire language because they're smart? Or are they smart because they have been able to acquire language? One popular theory of the evolution of the human mind supposes that our specialized adaptations to the human ecological niche carried with it, almost by accident as it were, an increase in general intelligence. A neocortex that expanded to handle visual processing and the control of precise hand motions and the sequencing of complex manual tasks was a neocortex that had stuff left over to do a lot of other things, too. The fossil record of early man is a record of ever more precise hands and ever more refined tools. Two million years ago, *Homo habilis* was making stone scrapers and flakes, and already the apelike hand with short thumbs and curved fingers was giving way to the more modern human hand with its long, fully opposable thumb and straight, fine fingers. His brain was still half the size of a modern human's. So did the brain generally expand to meet the demands of this ecological adaptation to tool making, and is that what made us so smart?

The more provocative notion, and one that may well fit the evidence better, however, is that our general intelligence is not very different from that of any other mammal's or bird's; the only thing that really distinguishes us is the species-specific adaptation of a language-acquisition device.

By all appearances, language certainly seems to be just as hardwired an attribute in humans as navigation is in pigeons. Children do not need to be taught language; even deaf children show a spontaneous mastery of grammatical concepts. Chimpanzees can be trained to within an inch of their lives and bribed with M&Ms to master a few dozen learned associations between symbols and things they want; humans without any formal teaching at all learn thousands of words and the rules for combining them in an infinite number of novel combinations to express ideas. In humans speech is controlled in a specialized region of the cortex called Broca's area; humans with damage to this region from an accident or stroke have difficulty talking. Understanding appears to reside in a second region, Wernicke's area. You can cut the corresponding parts out of a monkey's brain without affecting its ability to produce or respond to vocal calls.

That humans have a hard-wired sense of grammatical structure was strikingly supported by a fascinating study in which human subjects were asked to memorize unpronounceable strings of letters, like PVPXVPS and TSSXXVPS and PVV. The strings were three to eight letters in length, and subjects were given four at a time to learn. Unbeknownst to them, however, the strings had been generated in accordance with "grammatical" rules that, as in a real language, specified how letters can be replaced with other letters in a word. A second, control group was given strings made up of the same letters of the same lengths, but the letters were strung together in a purely random order. The group that got the grammatical gobbledygook learned new sets of words at an accelerating rate that significantly outpaced the control group that had to struggle with merely random gobbledygook. Perhaps even more interesting was the fact that when the "grammatical" group was told that their strings had been generated according to rules (but not what those rules were), they were able to do a quite good job of telling grammatical words from nongrammatical words—even as they themselves were unable to explain what had guided their choices.

When we try to zero in on the most basic *nonverbal* cognitive abilities of humans, we find that people do them efficiently, quickly, and in no ways strikingly different from, or better than, nonhuman primates, cats, rats, or pigeons do them. We can remember seven or so items in a list at a time; we can orient ourselves in a room or a landscape; we can tell at a glance whether one pile of M&Ms has more than another; we can keep our feet moving the right way on uneven ground; we can tell if a rotated object matches an unrotated one; we can recognize familiar faces; we can tell any moose from any car at a glance. And so can apes and monkeys and pigeons. Given the similarities in the cognitive abilities of those very different nonhuman species, it would be a dualist stance par excellence to try to argue that humans perform these basic, nonverbal cognitive tasks in a way fundamentally different from all other animals in creation.

Language is something else entirely. It is extremely significant that when we use language to tackle problems we are slower than computers, and slower than the innate, *nonverbal* mental processes our brains perform effortlessly and mostly unconsciously. The basic wiring of all brains, as the early artificial intelligence researchers found to their great frustration, did not conform to that of a general-purpose digital computer that manipulates data in sequential, logical operations. But a brain plus language does—if it is admittedly a slow and rather plodding general-purpose problem solver.

The key is that language is a system that allows ideas to be represented ad infinitum: Language automatically offers the means to represent ideas about ideas about ideas about ideas, and so on as far as we want. Language may well be *the* thing that makes possible the all-important leap from merely having intentions and beliefs to having intentions and beliefs about intentions and beliefs.

Language in other words *is* a discontinuity. The discontinuity between our minds and the minds of other animals is thus not a matter of degree or quantity—it is not that we just have more stuff, more memory, or a faster central processor unit. It is rather a matter of quality: We have a terrific piece of software that they simply do not. The peculiar fact of this piece of software is that it is a program that allows symbols to represent and manipulate other symbols on a completely different order from what the hardware by itself does. While a pigeon has a routine that performs a specialized application, language is a routine that performs a *general* task.

Being able not just to think, but to think about thoughts—to record them, reflect on them, experiment with rearranging them—is a step of huge adaptive significance. It allowed the first linguistic hominids to imagine solutions to problems before they tried them, to speculate about the thoughts of others (and to recognize the fact that there are such things as thoughts in the first place!), to design tools, to plan hunts, to share information, to teach one another without limitation (anything you can imagine in words you can explain to someone else in words). Language is what ultimately allowed humans, and not all of those clever apes and carnivores and bees and pigeons, to develop physics and astronomy and philosophy and ethics and electronic engineering and linguistics and history and education and justice and neurophysiology.

It is tempting to see in language an explanation for consciousness, too; language provides an automatic means for representing thoughts; when we are aware of our thoughts the thing that we are aware of is thoughts expressed in language. The artificial-grammar experiment (the one with the "words" like PVPXVPS) suggests that the acquisition of language itself is mediated by unconscious processes; that unconscious bootstrap hoists language, and thus consciousness, into place.

Yet that may be a stretch. There is evidence that human infants, even before they acquire language, begin to form concepts of mental states in others (see Chapter 6). Autistic children may lack an ability to attribute thoughts to others yet have no difficulty mastering language. So while language gives us a powerful tool for representing mental states in ourselves and in others, that may not be sufficient in itself to explain consciousness. And there is the point, too, that many of the "gut feeling" thoughts and conclusions we reach about the motivations and honesty of others seem to come from a deeper, nonverbal part of our minds. Yet the tight similarity between the meta-representation inherent in language and the meta-representations that are a necessary condition for consciousness and mental-state attribution are impossible to deny. Perhaps there was a coevolution of the hardware for one and the hardware for the other.

In our nonverbal and unconscious thoughts we may be remarkably similar to the animals with whom we feel such kindred feelings. It is more than just anthropomorphism that gives us a sense of connection with the animals of the world. The discontinuities that divide us are less a matter of biology than a matter of what one might almost call the super-biological phenomenon of language that our minds uniquely generate. Language is something that transcends the special-purpose hardware of the minds of man and animals. It is a universal algorithm that transcends the special-purpose algorithms that are tied to more specific tasks. The conclusion is in a sense one we have known all along: We are both the same and profoundly different.

2

THE SCIENCE OF HOW DO
WE KNOW FOR SURE

HAMADYRUS BABOONS LIVE IN HAREMS. A SINGLE MALE controls several females, and he spends a good bit of his time looking out to make sure that none of them tries to mate with any stray males in the area. But one day a young female was observed to do this: she left her male leader, hid behind a rock, and repeatedly mated with a young male baboon. Between these illicit rendezvous, she would return to the troop, approach the leader, and present herself to him—then back behind the rock for another brief encounter. Female baboons usually make a lot of noise when copulating, but in this case she was silent.

Male fireflies spend almost all of their nights trying to find females to inseminate. Males of the small species *Photinus pyralis*, common in the eastern United States, begin to search grassy areas shortly after sunset; every six seconds or so they emit a half-second flash while flying in a J-shaped swoop and wait for a female to respond with her own half-second flash. After mating, the female heads for a burrow to lay its eggs while the male resumes its search for available partners—which means there are always many more males than females out there hunting for mates, and the competition among males is fierce. (A study that followed 199 individual males found that their 8,000 flashes produced a total of two female replies one night.) But sometimes their ardor leads them to their death: A male's flash is met by an answering female. As he

21

approaches, the female, perched atop a tall blade of grass, begins to move down the stem to a more concealed position, flashing more dimly and less frequently. When the male lands and approaches his partner-to-be, the female—who turns out to be not a suitable mate at all, but rather a large specimen of an entirely different firefly genus, *Photuris*—pounces on him and eats him.

Like baboons, roosters live in harems. Like baboons, they are always working hard to keep their females in tow. When roosters find food, they emit food calls, especially when females are nearby, and the hens respond by approaching and sharing the food. Particularly good food gets a particularly long call. But in the presence of unfamiliar hens, rooster show some unfamiliar behavior. As an experiment, roosters were given either a highly desirable food (a mealworm) or an item so undesirable it wasn't even really food (a nutshell). When a rooster was by itself or in the presence of another male, it almost never emitted a call when it was given a nutshell. In the presence of a familiar hen, a nutshell would elicit a food call 17 percent of the time. But when a strange hen was introduced onto the scene, the rooster would emit a food call 50 percent of the time when it was given a nutshell.

DECEPTIVE DECEPTION

All three of these tragicomic episodes have been cited in recent years as remarkable examples of *deception*—and, more generally, as evidence of animal intelligence that involves insight, conscious thought, and understanding. Donald Griffin calls the fireflies' complex patterns of false signaling a "possibly rational, even Machiavellian" ploy and suggests that the "traditional" assumption that insects behave in a rigid, genetically preprogrammed manner "may well be a serious oversimplification." The deceptive rooster who pretends he has food in order to lure a new female into his grasp has been attributed with an "intention to misinform." The baboons that ducked behind the rock have been cited by many authors as an example of an extremely sophisticated feat of understanding; the young male had to realize that the male leader could not see him from that position, the female likewise had to know that by keeping silent the leader would be none the wiser (a feat of deception via "acoustic hiding"). Not only did they need to plan ahead and carry out a course of action; they had to grasp that another baboon possessed knowledge different from their own; they were able to figure out what the leader knew and did not know, and act accordingly. They showed that not only did they have their own intentions, beliefs, and desires, but that they could readily impute intentions, beliefs, and desires to others. They did not merely think; they possessed a "theory of mind."

Or did they? One of the reasons common sense is such an unreliable guide to the interpretation of seemingly thoughtful and intentional animal behavior is that the world is full of what cognitive scientist Daniel Dennett aptly calls "unthinking intelligence." The study of animal cognition might be defined as the science of How Do We Know for Sure, for learning and evolution do a sensational job of generating intelligence without conscious intention or insight. So, every time we encounter an example of intelligence that we are inclined to attribute to thinking, we had better be awfully sure that it is not just another instance of the unconscious brilliance of nature.

In the last chapter we encountered a number of examples of seemingly intentional and conscious behavior that, on closer examination, proved to be the product of simple associative learning. The dog learns that if he jumps up on the door and scratches, the door eventually opens. One thing follows another, and the connection is reinforced, even though no insight or understanding of the underlying mechanism is involved. When the one thing was actually *caused* by the other, the learning is nonetheless "intelligent." A horse learns to associate buckets with food, or someone heading toward him in the field carrying a halter and lead rope as something to be avoided.

Learning is a brilliantly efficient *general* paradigm for dealing with the endless *specific* complexities of the world. At its simplest, learning is nothing more than a mental linking together of an animal's own action, or a stimulus from the environment, with another stimulus that occurs at more or less the same time. Yet things that happen together usually *are* linked by cause and effect. A hunting tactic that fails a cat usually fails because it was a bad idea. A place where a zebra finds water will usually have water again. A dog that goes after a skunk and gets sprayed really will get sprayed again if he tries it again. It is convenient that the world is generally a rational place, for it makes the exquisitely simple strategy of learning by rote so effective and "intelligent" in its results. Even the conservatism of the learning process is part of what makes its outcome so "intelligent." Animals often seem to insist on doing things exactly the same way once they have learned to do so; they also may take many repetitions of an experience to forge a learned association. A lot of this can seem to us in the short run a manifestation of stupidity, not intelligence. The cat that keeps looking in the old spot for its food when its bowl is moved a few feet away strikes us as having rewritten the definition of "knucklehead." But being slow to learn and slow to change is generally an effective strategy, because the world, if not an irrational place, is at least a place full of ephemeral phenomena; waiting for something to be repeated is a way to filter out coincidences.

Proponents of a liberally mentalist interpretation of animal feats complain that attempting to explain these accomplishments as simple learned associations often takes us down a tortured and contrived path. But "simple" learning is such a generalizable tool that its scope and power is more considerable than people usually are ready to give it credit for. There is nothing to say that the deceiving baboon lovers do not have a theory of mind and a conscious intent. But simple rote learning is a straightforward and perfectly plausible alternative. A young male that attempts to approach a female gets walloped by the harem leader. (Learned rule: don't try to copulate when you see the harem leader around.) The young male baboon thus need not have had a theory of mind at all; he just picked a spot where *he* could not see the harem leader. Similarly for the female who suppressed the usual copulatory vocalizations. Simple learning could have taught her that making sounds brings the harem leader, and trouble. My dog has clearly learned that his own barking brings a person over to open the door so he can come in. If, rather than opening the door, I clobbered him every time I came over in response to his barks, it would not take long for him to stop barking. Would we call that "deception"? Would we say that my dog is deliberately trying to conceal from me the fact that he really wants to come in?

Or consider this sneaky behavior of my old collie: There is a phone down in the tack room of my barn, which is also where the cats are fed. We usually keep the door shut so the dogs cannot get at the cat food, but sometimes when I go in there to fetch something the collie will come in and start eating the food. When she does, of course, she gets yelled at. Then she hit on a new strategy. She would never eat the food when she came into the room with me—except when the phone rang and I started talking to someone. Then she would give her full attention to the cat food, and usually get away with it.

Now one explanation is that my dog consciously calculated that I was distracted, that she was aware that when I was speaking on the telephone I had a knowledge of her behavior that was different from her own. The actually much simpler, and not at all tortured, explanation is that she was always trying to get the cat food whenever she was in the room; trying it once when I was on the phone formed a simple learned association: eat it when the guy is holding the phone and talking.

Dogs offer many wonderful illustrations of clever behavior which are fertile ground for speculation about "unthinking intelligence" versus understanding. The fact is that even the most complex and insightful behavior of dogs can almost always be readily explained simply, and without torturing the case, through associative learning. Again, this doesn't prove that dogs *don't* have conscious understanding; it only cautions us that we cannot jump to the conclusion that they do. Few dog owners, for example, would interpret their dogs' cringing greeting after they had messed on the carpet as anything but an expression of conscious guilt. But consider this: all dogs are house-trained by being taught to associate their act of defecation—or the presence of dog poop on the floor and ourselves on the scene—with punishment. So the condition the dog has to learn in order to display what we take for guilt is actually quite simple: dog poop in the house plus a person equals trouble. As social animals, dogs as a matter of course and quite instinctively display a variety of behaviors of submission—cringing in particular—toward an animal higher in the pack hierarchy (in this case, us) that displays aggression. So is it guilt, or is it merely appeasement of aggression that the dog has been conditioned to anticipate under these circumstances? My dog, after a few accidents, did not particularly display anticipatory "guilt," but quickly learned to associate the presence of a person cleaning up the floor, and holding a newspaper with dog poop in it, with trouble. He once ran upstairs quite happily to see my wife, who was in the attic cleaning up his mess, but made a U-turn and took off back down the stairs as soon as he saw the poop-laden newspaper she was holding.

INTELLIGENCE AT LARGE

If all of this seems a stretch, then just consider the remarkable number of complex "superstitious" associations that dogs readily acquire. In his book *The Farmer's Dog*, John Holmes tells of one of his dogs that once collided with another dog as they were both running through an open gate at high speed. After that, the dog would still go through the gate quite willingly by himself, but refused to do so whenever the other dog—or even a third dog who was present on the occasion of the wreck, though not a participant in it—was anywhere near. When a behavior that is arguably nothing more than the product of associative learning reflects an accurate generalization of the underlying reality, we readily interpret it as clever insight; how, then, do we interpret phenomenal stupidity of an association that reflects an inaccurate generalization of coincidence?

Superstitious behavior is remarkably common in domestic animals. A horse will repeatedly shy at the same spot on the road where a bird once flew out of the bushes and startled him. Dogs will take a longer and more circuitous but habitual route to the mailbox. Often such superstitious, or simply odd, behavior is self-generated. An animal happens to do something that is followed by a favorable consequence, and the link is forged. I let my Border collie out of the house every single morning, but he has a whole series of rituals that he performs every single time, having come to associate these actions on his part with the subsequent opening of the door: One is that he waits until *after* my wife or I get out of bed, and then lets out a peculiar howl-bark that he uses at no other time. You or I may say this is ridiculous; it is not as if he needs to tell me to let him out, since I do it every morning. But from his point of view this association has been perfectly reinforced. Every morning the following things happen: I get up, he makes a funny noise, then I go downstairs and open the door. Much like the man in Nebraska who was always tearing scraps of paper and throwing them on the ground to keep tigers away, it works every time. Similarly, one of my old collies once came up with the most bizarre performance of backing into our other dog while making a strange whining sound. There was absolutely no apparent reason for all of this (in fact, the other dog often bit her on the nose for disturbing him)—except that it always got a lot of attention as her human audience laughed and petted her when she was through.

Or consider the case of the dog "smile." Wild dogs do not open their mouths and show their teeth unless they are being threatening. Domestic dogs do it all the time in their friendly interactions with us. The mentalist explanation is that they are imitating our behavior. In other words they see what we do, consider what actions on their part would produce the same visual effect to an observer, and decide to copy it. One problem with this explanation is that controlled experiments of learning through imitation have been overwhelmingly negative in the case of nonprimates. Dogs and horses, for example, simply do not learn to perform an action any faster by being able to see a person or a conspecific perform it first. (A horse will more readily perform an action such as jumping over a stream if another horse leads the way, but this is merely the herd instinct, not learning.) In this case the learning explanation is again not only perfectly plausible but also much simpler: Consciously or unconsciously, we are rewarding our dog with attention when it "smiles," because it is an appealing facial expression to us.

The intelligence of learned associations of this sort lies in part *outside* the animal's brain. It is not what is inside the head, but what the head is inside of—to use William Mace's felicitous phrase from an essay advocating an "ecological approach" to the study of animal intelligence. It is the rationality of the world, the fact that there is more often than not a cause and effect between associated events, that makes a simple process of association "intelligent." The very reason domestic animals appear so smart is that a lot of the intelligence of their actions and responses actually resides in *our* behavior. A dog that has learned to stand by the door to the closet where its leash is kept when it wants to go out does so because of our ability to read such signals and reinforce them by our response. (The intelligence of the answers we get back from the person in Searle's "Chinese room" that we encountered in the introduction is the sum of his mind's ability to follow rules plus the rules themselves that have been supplied to him to follow. The intelligence resides in the sum total of stuff *in the room*, in other words, not just in the person's mind.) One of my Border collie's more subtle attention-getting tricks involves some perfectly timed barking when my wife is on the phone. Like most dogs he is ever seeking attention; if he tries to do so by barking or otherwise making a nuisance of himself while my wife is on the phone she shoves him away and ignores him. But he has a remarkable ability to pick up on the shift in intonation in one's voice (something I for one only consciously noticed once my dog's behavior pointed it out to me) as one approaches the end of a phone conversation. It is difficult to describe intonations in words, but you can notice it yourself if you listen to someone as they say "Okay, thanks for calling, good to talk to you, talk to you soon, take care now," and so on. Whenever my wife reaches that point in her call, the Border collie starts barking and nosing her. He has obviously made an association between those sounds and her subsequent hanging up the phone and renewed attention to the dog, an association reinforced by the fact that my wife usually gives up trying to shut him up when she is through with the substance of her conversation and is about to hang up anyway. But the point is that the subtle cleverness of this behavior reflects two things: one, the dog's very acute sensitivity to vocal signals and his ability to form associations and, two, the complexity of our human social structure which provides so many contextual cues to our own rational behavior—which give the dog such a rich field on which to exercise his learned associations.

EVOLUTION IS SO SMART

Plants are surely dumber than fireflies. Yet, just as unthinking learned associations usually capture (unthinkingly) an intelligent appreciation of cause and effect, so has evolution, even more so, demonstrated a genius for manufacturing the functional equivalence of intelligence in even the simplest organisms. As Daniel Dennett has pointed out so well, natural selection behaves precisely as if it *were* guided by conscious intention—and very often an intention to trick, outwit, or confuse other organisms. Carnivorous plants give off a smell like that of rotting flesh to lure flies to their death. Fruiting plants offer a tempting lure for a different end, to trick mammals and birds into dispersing their seeds, and provide free fertilizer in the process. Some plants refine this stratagem further by making the seeds of their fruits bitter; bitter flavors are unpalatable to mammals such as mice, which chew up the seeds and destroy them, but not to birds, which swallow and defecate the seeds whole. Weeds show a diabolical cleverness in imitating the growing habits of harvested crops so they get a free ride. Some are indistinguishable from the crops whose fields they invade until they reach the flowering stage, by which point they have already been fertilized, cultivated, and tended. Others produce seed that can scarcely be told from the seed of the harvested crop that is saved to be planted the next year. Clovers produce a chemical that mimics estrogen, suppressing ovulation in sheep that graze on them—the ultimate revenge, a plant that gets back at its predators with a birth-control pill.

Viruses are the ultimate example of diabolical intelligence that cannot possibly have intent. Viruses consist of nothing but a strand of DNA or RNA in a protein coat. It is hard to imagine evolution in a more basic form. Consider what the rabies virus does: it attacks the nerve cells of the brain in just such a fashion that the victim has an urge to bite other animals—which is just about the only way the virus can spread itself to other victims. The rabies virus understands neurophysiology better than a neurophysiologist. Or consider what the AIDS virus does: it attacks the very cells of the immune system that the body uses to fight off viral infection. It understands immunology better than an immunologist.

The physical forms animals have evolved often embody deception and trickery. Butterflies have large spots on their wings that resemble eyes, which startle predatory birds, often deflecting an attack. Stick bugs blend into their surroundings to elude predators. Roundworms that parasitize one shrimplike creature cause this intermediary host to turn blue, making it more conspicuous to the ducks that feed on it—providing the roundworm an assured entry into the gut of its final host, the duck.

We read of such accomplishments of evolution with astonishment. These are not merely clever strategies—these are fiendishly clever strategies. We wish we could thinks up things like this. On the surface, every one of these stratagems would seem impossible to develop without a "theory of mind." To hit upon the idea of making a stinking flower to attract flies would seem to require knowing what it is like to be a fly. To produce a seed that looks just like the grass or clover that farmers carefully harvest and store and sell to lawn-growing homeowners all over the country would seem to assume a knowledge of twentieth-century American agricultural practices and landscape tastes. But of course all natural selection has done is throw out the things that do not work and keep the things that do. A blind, unthinking, trial-and-error process produces some of the smartest things we know of.

Behaviors that are intelligent, crafty, and deceptive are no less common in nature than are physical and chemical stratagems. All predators stalk. This is surely a brilliant strategy, and a theory of mind explanation would be quite elaborate: the predator apparently has to be aware of the sound he himself makes as he moves, the possibility that a prey animal can have a state of knowledge different from himself about his movements, and that stalking is a way of muffling his sounds. "Acoustic hiding" surely. But puppies a few weeks old start stalking bugs (and one another), and by all appearances it is an instinct as automatic as eating or running. Any ancestral carnivore that didn't stalk would not have been around very long to pass on its nonstalking genes.

In an aggressive encounter a dog raises its hackles. This too appears to be a purely instinctive behavior, and it is almost certainly a ritualized signal derived from the fact that big things look intimidating. A dog raises its hackles "in order" to look big. But again no conscious intent is necessary to explain this—just evolution. Dogs raise their hackles in aggressive encounters because natural selection favored dogs that did so. It works.

Some of the most elaborate behavioral ruses involve the routines prey animals use to distract or fool predators. Opossums play dead when attacked, a particularly brilliant strategy because it is perfectly tuned to faking out most predators' instincts. Piping plovers lead predators away from their ground nests by pretending to have a broken wing and running off in a misleading direction.

Fireflies that mimic the flashes of other species are just one among thousands of examples of specialized behaviors that evolution has crafted in the course of the evolutionary "arms race" between species. Donald Griffin nonetheless insists that conscious intent could well be at work here. He argues that it is inconsistent to accept the unobservable phenomenon of evolutionary selection as an explanation for the adaptive, intelligent, and deceptive behavior we see in nature while rejecting the no more unobservable phenomenon of animals' mental states as an explanation. Moreover, he and others suggest that the very flexibility of the fireflies' deceptive flashes or the predators' stalk or the plovers' broken wing displays proves that they are far more than mechanical reflexes. Predators modify their stalk depending on circumstances. Fireflies seem to change deceptive strategies as needed, coming up with elaborate variations on the theme: Males of the predatory species trying to find a mate sometimes flash like the males of the prey species that the females of their own species are trying to lure in—faking out a mimicry with a mimicry of their own. The females, trying to look like a mate to get a meal, are deceived by males trying to look like a meal to get a mate.

A study of plovers found that the birds generally watch the behavior of the intruder and appear to shape their display accordingly. If the intruder ignores the display, the bird will often repeat it more emphatically from a new position. When followed, the birds regulate their speed so they are always just temptingly out of reach. They also appear able to quickly learn to distinguish between more and less threatening intruders, ignoring herbivores but putting on their act for carnivores. They react more strongly to people who had previously walked within a few feet of the nest than to those who had passed at a distance.

From this Griffin—and, to a lesser extent, Carolyn Ristau, who conducted the plover experiments—conclude that the birds' behavior shows actual intention to deceive. Griffin say that "adaptability to changing circumstances" is evidence of "conscious thinking."

But there is something wrong with this argument—several things, in fact. First, it is not as if plovers, as a predator approaches, sometimes fake a broken wing and other times dance the Charleston. All predators stalk, all opossums play dead, all plovers fake broken wings. No other explanation answers but that these are innate, genetically programmed instincts, honed by evolution because they work.

The variations on these themes that animals display do, however, suggest some degree of computation. And it is here that most of those who favor the cognitive approach part company with Griffin. The whole point of the cognitivists—and this is the second problem with Griffin's argument—is that there is a vast middle ground between the two simple choices of animals' either being conscious, thinking, intentional beings, or mere assemblages of mechanical reflexes directed rigidly by their genes. "It doesn't take a lot of neurons to be deceptive," as biologist Roy Caldwell of the University of California–Berkeley has said. Even complex, flexible, adaptable, intelligent behaviors are often simple enough in their underlying principles of operation that they can be readily simulated with quite simple computer programs—without any "understanding." Decision making does not in and of itself imply consciousness. If that were the case, we would be routinely compelled to attribute consciousness to bacteria and plants. Venus flytraps close up only when they have trapped a fly; if what they have trapped turns out to be a twig, they open again quickly. Bacteria move toward a high concentration of a chemical substance they are "looking" for.

Nor, importantly, does the fact that a behavior is the product of evolutionary adaptation mean that it has to be rigid, inflexible, and stereotyped. As animal behaviorist David McFarland notes, "We can expect natural selection to shape the decision-making mechanisms of animals in such a way that the resultant behavior sequences tend to be optimally adapted to the current situation."

To summarize: evolutionary selection, associative learning, and true insight can all produce the same intelligent behaviors. Evolution is trial and error on a very long time scale. New solutions are randomly generated; the ones that don't work are rejected—through the death of the individual. Learning is much the same process but on a much faster scale. Possible solutions are tried; those that don't work are rejected by the failure to form a learned association. Insight, as Daniel Dennett has pointed out, is in a sense the same process as well, only that here we try out the solutions in our minds first. As the philosopher of science Karl Popper said, this prescreening of possible solutions "permits our hypotheses to die in our stead." But when we look at nature, we are only looking at the survivors. We are seeing the end product of successful evolution and successful learning. No wonder it is so easy to confuse that with true understanding and insight.

MOCK ANTHROPOMORPHISM

Defenders of continuity between the mental lives of man and animals often invoke Darwin as the prime supporter of the cause. While Darwin introduced the now-unquestioned conclusion that all life on earth is linked by common ancestry, his theory of evolution in many ways undermined the extravagant mentalist claims about animals' behavior that his theory itself at first was taken to sanction. Before Darwin, the only way to explain purposive or intentional behavior was by imagining that an "informing spirit" was resident in the organism itself. But evolution's ability to imitate conscious intent changed everything. As the zoologist Gordon Gallup has observed, "organisms have evolved in many instances to act as if they had minds." Today, as John S. Kennedy points out, biologists use the word "purposive" merely as a metaphor for *adaptive*—in the strict evolutionary sense. Evolution is an incredibly powerful engine for honing purposiveness because adaptive strategies that did not work over the course of evolution did not survive. Darwin's breakthrough was not in fact a license for anthropomorphism, but on the contrary a caution against assuming that the organism must itself possess the intent or purpose that it displays.

On the other hand, precisely because evolution so often yields something indistinguishable from conscious purpose, we can learn a great deal about why animals do what they do if we pretend that evolution actually *does* have a conscious purpose, and see where that train of thought leads us. Asking what an animal is "trying to do" is a way to creatively probe the adaptive, evolutionary purpose for its behavior. Asking why the roundworm "wants" to turn its intermediate hosts blue can help us think up experiments that can get to the bottom of the adaptive point of it all. For example, thinking this way could lead us to test whether ducks more readily find blue things.

Kennedy calls this "mock anthropomorphism"; others have used the term "critical anthropomorphism"; Dennett refers to this heuristic approach as the "intentional stance." It works so long as we keep in mind that we are probing only natural selection, and possibly the cognitive processes natural selection has generated. We are not probing what's going on in the animal's mind—for, as Kennedy notes, it is natural selection that has assured that what an animal does "makes sense." We are drawing an analogy from the way our minds work—from our sense of intentionality—to the process of natural selection, not an analogy from our minds to animal minds.

This approach is the foundation of evolutionary ecology, a burgeoning field of research that has produced wonderful insights into the adaptive value of otherwise inexplicable behaviors and physical forms that appear in nature. This approach works as well for behavior as for physical form, as well for plants as animals. Darwin, for instance, observed that the flowers of the Madagascar star orchid had unusually long tubes. He then in effect asked, what would they want that for? Recent evidence has at last confirmed Darwin's speculation that a longer tube increased the chances that the moths that pollinate these flowers would bring their legs and bodies in full contact with the flowers' sexual organs as they feed on the nectar at the bottom of the tubes, thereby increasing the amount of seed set. But the studies also revealed an "arms race" between the flowers and the moths. Long tubes favor moths with long tongues so they can reach the nectar at all; long tongues in turn push the flowers toward even longer tubes to again force the moths to come into contact with the sexual organs at the top of the flower (that is, rather than just cheating by sticking their long tongues in from a "stand-off" position that doesn't bring their bodies into contact with the sexual organs). Explicitly using anthropomorphic language like "arms races" and "what the flower is trying to get the moth to do" is an enormous aid in figuring out what's going on here.

Another example is the curious observation made about the copulatory habits of dungflies. Males usually break off copulation before they have fertilized all of the female's eggs. Again, asking why they should want to do that led to an answer that was not only convincing but quantitatively verifiable. Suppose a male fly has an intentional goal of fertilizing the maximum number of eggs. It takes him a certain amount of time to search for and find a new mate. During a single copulation a point is reached where in effect he has to choose: do I stick with this one to fertilize the last few remaining eggs, or is it a better use of my time to cut out now and find a new mate and start from the beginning with her. From actual observations, researchers were able to observe how long it takes a fly to find a new mate and the rate at which eggs are fertilized during copulation, and calculated that each copulation should last about 41 minutes and result in about 85 percent of the eggs fertilized. The actual numbers were just slightly less.

"Nowadays, recognizing that natural selection is an optimizing force, we do not hesitate to hazard a guess as to what an animal will do, based simply upon what seems to us would be best for it to do; and it will often be a good guess," notes Kennedy. Bushmen in their hunting of animals are reported to use this approach all the time; they discuss what they would do if they were the animal, and their conclusions are said to be "amazingly accurate." The animal rights advocate Michael W. Fox cites this as evidence of animal consciousness, and complains that scientists in their anti-anthropomorphic zeal reject such proofs.

But of course that entirely misses the point about natural selection. What scientists are rejecting is the "genuine" anthropomorphism that equates apparent purpose with conscious intent.

The confusion between the two sorts of anthropomorphism that Fox's and Griffin's arguments reveal is the reason that even "mock anthropomorphism" is full of dangers; it is all too easy to shift from using a phrase like "the fly wants to figure out how to spread its sperm as efficiently as possible" as a purely evolutionary metaphor and to start, even unconsciously, taking it as a statement of conscious intent.

Gordon Burghardt of the University of Tennessee underscored this danger in an article that called for critical anthropomorphism as a tool to formulate testable hypotheses, but cautioned that, in his experience, "unless challenged to separate description from interpretation, students readily use and defend the use of sloppy teleological and anthropomorphic *thinking*, not just vocabulary." Teleological statements are those that take the result of an action as its intent, the apparent function of an action as the initiating purpose for the action. Even within the strict confines of evolutionary explanation, metaphors are always teetering on the edge of the teleological (the longer flower did not evolve "in order" to get moths to more efficiently fertilize them; that was an end result of natural selection—the better fertilized flowers were more likely to survive and pass on the trait that permitted more efficient fertilization). But when extended to the realm of animal behavior, the confusion is perhaps inevitable.

THE NATURE OF LYING

Besides what we might call the "gross" anthropomorphism of directly attributing human thoughts, intentions, and feelings to animals, there is a much more subtle anthropomorphism that the use of metaphorical terminology often manages to sneak into our thinking. The very terms that increasingly are being used by popularizers of animal behavior studies and indeed by researchers themselves in the scientific literature tend by their very nature to "up the value" of the behaviors so described.

Some of this is innocent, and perhaps inescapable. Colorful short-hand is the soul of vigorous prose. The "selfish gene" is a catchy phrase. Describing as "rape" the copulation between wild horses in which the stallion dispenses with the normal courtship ritual and forces a mare pregnant by another stallion to stand is a vivid (and far more concise) way of putting it. But it still has woeful consequences for interpreting the significance of the data.

The very use of term "food call" in the case of the mendacious chicken that opened this chapter is a case in point. Human speech has semantic significance. The actual sounds of speech are arbitrary; there is nothing about our evolution (except for the limitations it imposes by way of our tongue and mouth and vocal cords on the sounds we can produce) that requires us to use any one particular bunch of noises to stand for the concepts of "moose" or "squirrel." We could just as eas-ily have used the sound *skwûrəl* to mean "moose" and *mōōs* to mean "squirrel." Human language represents an encoding of a discrete idea by a sender and a decoding of that message by the receiver. Being only human, that is the way we're used to thinking about all kinds of com-munication, as a sort of code in which a well-defined quantity is repre-sented by a stand-in. The stand-in can be a series of dots and dashes, like Morse code; or the series of 1s and 0s that a computer uses to relay data to another machine; or the consonants and vowels of speech.

So when scientists began studying animal calls, almost without even thinking about it they started giving calls semantic labels. They would watch what the animal was doing at the time it emitted a call, and draw what seemed to be an obvious conclusion. A sound that an animal made when it saw a predator was an "alarm call." A sound it made during courtship was a "mating call." A sound it made when it found food was a "food call."

That very act of semantic labeling carried with it a huge anthropo-morphic assumption. We have set ourselves up for perceiving a "lie" by labeling one of the rooster's calls with a precise semantic meaning. It means "food": if the rooster uses the call in a circumstance where there is no food, it is obviously a deception. Right? Well, right as far as the assumptions we have made. But let's consider an entirely different scene for a minute. General Halftrack in the comic strip Beetle Bailey is always ogling his buxom secretary, Miss Buxley. You, an investigator from the adjutant general's office, have responded to an anonymous tip about the general's sexual harassment by installing a hidden camera in the general's office and are now sitting down to watch a week's worth of film. You know from Miss Buxley's job description that she is supposed to go into the general's office to file his paperwork whenever the general pushes a buzzer. On Monday, Tuesday, and Wednesday, the general pushes the buzzer at 11 a.m.; Miss Buxley comes in and does the filing; the general leers down her dress. On Thursday, the general pushes the buzzer at 11 a.m.; Miss Buxley comes in; the general leers down her dress and tells her there's no filing to be done. On Friday, Miss Buxley takes the day off; her substitute is an even more buxom clone of Miss Buxley; the general has no filing at all to be done that day but pushes his buzzer at 9, 10, 11, 12, 1, 2, 3, 4, and 5, ogling the temp each time.

The general's obviously a lecher. You've got him on thirteen counts of lascivious peeking. Even worse, you've got him on ten counts of deliberately misusing a government-issued buzzer to give false and misleading commands to a subordinate—he has obviously been tricking Miss Buxley and her substitute to come into his office just so he could look at them.

The general, an honest man, confesses to the charges of lechery at once. He adamantly denies the charge of lying. His defense is simple: *You* interpreted the buzzer to mean "it's time to do the filing." But I just use it to mean, "Miss Buxley, come here." He goes on to explain that if every time Miss Buxley came in he told her there was no filing, she would stop coming in. So most of the time he only buzzes when there really is a job to be done. In the case of the temp, though, he was naturally far more interested in getting a look at her, and since she kept coming in every time he buzzed, being new and expecting that the buzzer really meant something, he kept buzzing. But General Halftrack insists he has been perfectly consistent and honest in the way he uses his buzzer.

The fact is that all of our conclusions about the rooster's deception follow from our assumption that a food call literally means "there's food here for you to eat." But there is plenty of evidence to suggest it doesn't have any such precise semantic meaning at all. Roosters in fact have been observed to use "food calls" in many other circumstances. It could "mean" as little as "here I am." Or rather it literally does not *mean* anything. An identical call is used in situations where no food whatso-ever is involved, especially when members of a flock are separated and are seeking to reestablish contact. Like many birds, roosters practice *courtship feeding.* They offer some food to a female that approaches. So the "deception" scenario may simply be an entire misreading of the sit-uation. The rooster could simply be announcing his presence; from past experiences hens have learned that if they approach a male making such a call they get some choice morsel, and the males have learned that if they offer a choice morsel they have a better chance of having females come in response to their call—and copulate with them. If there is a new female on the scene, the male may well be more highly motivated to mate with her and attempt to add her to his flock—and will thus be more likely to call even without any food to offer her. A final twist is that courtship feeding often becomes highly ritualized in many species; sometimes the food is regularly replaced by a non-food item. Some species of emphid flies, which like spiders wrap their prey in silk, pre-sent objects that are merely non-nutritious fragments of prey; one species presents empty silk "balloons" containing no prey whatsoever. So perhaps even empty nutshells are merely part of ritualized courtship feeding.

Whether we call something "deception" depends heavily on the assumptions we make about the communicator's intent. It has been said that man developed language so he could lie more effectively, and there is little doubt that our personal familiarity with semantic deception has preconditioned us to find the same in animal communication, even when there may be none. This all-too-human tendency is perfectly cap-tured in the Jewish tale of the two traveling salesmen, competitors in the same line of trade, who run into each other at the train station:

"Hello, Moskowitz."

"Hello, Finkelstein."

"So—Where are you going?"

"Minsk."

Long pause.

"Listen, Moskowitz, when you say you're going to Minsk, you want me to think you're going to Pinsk. But I happen to *know* you really are going to Minsk. So why are you lying to me?!"

A final analogy to clarify the dangers of unwarranted semantic labeling of animal sounds: Let's say your dog barks in a particular fashion whenever a strange car pulls up to your house. You come out and join your dog to see who it is. Then one day your dog barks in that way, you go out and find your dog holding a frisbee in his mouth and no car. If you call the bark a "strange car bark," then your dog is a devious liar. If you simply observe that your dog has discovered that whenever he barks that way a person appears, then it something much more mundane.

This practice of giving semantic labels to vocalizations has added no end of confusion to studies of animal communication, deceptive and otherwise. One important principle, as we shall see, is that many animal vocalizations are phenomena that have much more do with acoustics than semantics. For a long time people who studied horse vocalizations tried to give semantic meanings to the whinny with contradictory and inconsistent results. Horses whinny when they catch sight or smell of a familiar horse at a distance; they whinny when the rest of the herd has moved off and they have been left behind; mares and foals whinny to each other when separated; feral horses whinny to establish their location relative to other bands; stabled horses will sometimes whinny when they are very hungry and hear their owner approaching. A whole series of semantic labels would seem to be called for. A whinny means "come here" or "wait for me" or "I'm lost" or "stay out of my way" or "I'm hungry." But the essential characteristic of the whinny is that it is a signal that, by virtue of its acoustic properties, carries an exceptionally long distance through various sorts of environments. (Particularly significant is that a whinny starts high and falls to about half its starting pitch, sweeping a whole spectrum of frequencies; low buzzy sounds travel best in open environments, while higher pitches travel best in forests.) So the whinny is really is just a sound that is used whenever a horse needs to announce its presence over a distance. The "semantics" of the signal are really provided by its context—and not by the signal at all.

MOVING GOALPOSTS

A single word can freight animal behavior with a whole raft of assumptions. As noted above, some of this is the inescapable tendencies of our inherently anthropomorphic language. But some of this is less innocent. As Jean-Marie Vidal and Jacques Vauclair have noted, the "upping of the ante" is apparent everywhere these days. Animal vocalizations have become "symbols" and "signs" and even "language"; interactions between two animals are "exchanges"; an instinctive inhibition against consanguineous mating is an "incest ban"; the concealment of information is a "lie"; an ability to recognize other individuals of the group and their rank in the social hierarchy is "respect for others"; the attachment between mothers and young is "love"; involuntary copulation is "rape"; an act of assistance to another member of a group is "altruism" or even "moral sense"; the transmission of know-how is "teaching"; regular habits are "social rules"; rare behaviors are "transgressions" or "infractions" of those rules; defensive behavior or counterattack is "revenge" or even "just war."

Much of this is driven by a more or less absurd exercise that goes like this: Team A tries to come up with a list of defining characteristics that distinguish humans from other animals. Team B finds an exception to one or more item on the list. Team A comes up with a refined list. Team B finds another exception. And so on. For example, it was long said that humans are the only species that uses tools. Then it was pointed out that elephants use sticks which they hold in their trunks to scratch themselves, that sea otters place stones on their chests and use them as anvils to break open crab and mussel shells, that Egyptian vultures pick up rocks and toss them at ostrich nests to break open their eggs. Then it was said that humans are the only tool*makers*. Then it was observed that chimpanzees strip off the leaves of sticks or prepare grass stems to use as probes for fishing out ants or termites from their holes.

Then it was said that humans are the only species that uses a tool to make another tool.

Then Team B accuses Team A of moving the goalposts. But perhaps Team A should not have agreed to play such a ridiculous game in the first place. It should be obvious that the difference between man and other animals does not lie in a few categorical definitions, but in a sum total of specifics. To name the merest fraction: Humans can think about whether to go to the movies next weekend. They can discuss the meaning of a word. They can play a sonata on the piano. They can design a vegetable garden on a piece of paper, then plant it. They can ride a horse. They can calculate when the next lunar eclipse will occur. They can imagine the conversation they would like to have had with Groucho Marx. They can learn what happened in their country two hundred years ago by reading a book. They can fix an automobile engine. They can train a dog to retrieve objects. They can build a desk out of wood. They can debate a political or moral issue. They can confide their thoughts to a diary. They can plan a trip by studying a map. They can exchange humorous insults. They can write a list of things that people can do that animals cannot.

Yet invariably Team B presents its discovery of a crushing exception to the General Rules of Human Uniqueness as an explicit refutation of man's claims to special status, as a narrowing of our self-declared gap between us and them. Popular accounts of such research invariably cast it in these terms; the finding of "mathematical" or "language" or "tool-making" abilities in apes is described as further proof that the distance between humans and our "closest living relatives" has been narrowed once again. But equating what a chimpanzee does with a twig to the entire range of tool use in humans ought to be self-evidently absurd. To claim that because chimpanzees can tell the difference between three and four objects they share with man an ability to do mathematics is a triumph of superficial semantics over substance. Surely it does not take much thought to recognize the difference of essential quality between mastering calculus or trigonometry on the one hand and assessing the relative size of two piles of M&Ms on the other. Steven Pinker offers a delightful analogy in his book *The Language Instinct,* when he asks us to imagine what would happen if some animal behaviorists were elephants. Elephants are the only living animals that possess trunks, remarkable organs that are six feet long, contain 60,000 muscles, and enable their owners to carry entire huge trees. The elephant's closest relative is the hyrax, a guinea-pig-like mammal. Nonetheless, one school of the elephant animal behaviorists, Pinker writes, "might try to think up ways to narrow the gap. They would first point out that the elephant and the hyrax share about 90 percent of their DNA and thus could not be all that different. They might say that the trunk must not be as complex as everyone thought; perhaps the number of muscles has been miscounted. They might further note that the hyrax really does have a trunk, but somehow it has been overlooked; after all, the hyrax does have nostrils. Though their attempts to train hyraxes to pick up objects with their nostrils have failed, some might trumpet their success at training hyraxes to push toothpicks around with their tongues, noting that stacking tree trunks . . . differs from it only in degree."

One point, as Pinker notes, is that (as Stalin is reputed to have said about military force) quantity has a quality all its own. One can argue that *all* differences are just matters of degree, after all. An oak differs from an acorn only by degree; a man differs from a fetus only by degree; for that matter a man differs from an amoeba only by degree. At some point, though, a difference in degree becomes a difference of essential quality.

His other, more important point is that the "closest living relative" argument is nothing but a recapitulation of the old evolution-as-a-straight-line fallacy, which we encountered in chapter I. Evolution is not a straight line. It is an ever-branching tree. Two species that happen to be the most closely "related" to one another are related only by a very remote common ancestor. Most species have evolved independently for millions of years from their "nearest living relative," adapting to very different habitats and circumstances. Thus it is really just a matter of history and the vicissitudes of evolution and extinction that determine which species *is* our nearest living relative.

Yet we have never been quite able to shake this popular and wrong notion of evolution as a continual line of "progress" linking the currently living species of the earth, from amoebas to snakes to cats to apes to us. That evolution-as-a-chain-of-progress idea has saturated popular culture. Edgar Rice Burroughs' 1914 novel *Tarzan of the Apes* stars a band of pop-Darwinian apes that come off just as sort of uncouth humans whose brutish tendencies now and again get the better of them. Burroughs pedantically portrays his ape society possessing primitive or embryonic forms of every human institution—religion, courts of law, politics. Tarzan, for his part, reveals the apelike qualities that are latent in man, if concealed by civilization or atrophied by modern disuse. He has a great sense of smell.

Of course we did *not* descend from apes; apes and humans are linked only by a common ancestor that lived 7 million years ago. If we could bring the fossilized bones of early humans from Olduvai Gorge back to life, then indeed we would have material to probe for evidence of evolutionary continuity with us, for then and only then would we be seeing evolution in progress. But to hunt for humanlike characteristics in apes is a supremely self-centered occupation on our part, and one that really defies evolutionary sense. Suppose all the great apes had been wiped out, Pinker asks—would we then feel we had to see a mirror of ourselves in the monkeys? Suppose monkeys had been wiped out; would we then desperately look for signs of tool use and language ability in lemurs? Or, if the only species left on the planet were humans and amoebas, all intermediate forms and common ancestors having been wiped out by a comet that hit the earth, would we spend time pointing out the similarity in the DNA of our two species and looking for other signs of how closely related our two species are? Would we argue that Darwin's theory of evolution requires continuity between man and amoeba, and insist that those who deny the sense of that are flying in the face of Darwin himself?

As the linguist Noam Chomsky said recently in exasperation over experiments attempting to teach humanlike language to apes, "If you want to find out about an organism you study what it's good at. If you want to study humans you study language. If you want to study pigeons you study their homing instinct. Every biologist knows this." Sara Shettleworth, a psychologist at the University of Toronto, contrasts the "anthropocentric program" of cognitive research with an "ecological program." She notes that anthropocentric questions such as, "Can any nonhuman species count?" lead to a search for *demonstrations*, not *understanding*. It leads us to ignore the most "cognitively rich" behaviors of animals in favor of ones chosen purely because they resemble *our* abilities. And it leads us to totally neglect "consideration of what cognitive processes animals might be expected to have evolved for dealing with their natural environments." Certain birds, for example, such as Clark's nutcrackers, are very good at remembering where they have cached food. If we really want to understand animal cognition, and to place that in its true ecological and evolutionary context, Shettleworth says, study how nutcrackers accomplish this—not whether they can count the way humans do.

Trying so hard to breach the wall between man and other animals, we seem to have forgotten what we were fighting for. Herbert Terrace, once an enthusiastic ape-language researcher but now one of its sharpest critics, points out that finding trivial counterexamples to general criteria which supposedly define the line between man and beast really does miss the story. How do chimpanzees solve problems and make decisions without language? How do they remember things without language? "Those are much more interesting questions," he says, "than trying to reproduce a few tidbits of language from a chimpanzee [who is] trying to get rewards."

3

THE MIND'S SOFTWARE

🗲🗲 AS EARLY AS 1912, PIONEERING RESEARCHERS IN THE
nascent science of experimental animal behavior ran into some phenomena that could not fully be explained by simple learning. Köhler's
insightful apes, to be sure, turned out to be more an instance of wishful human thinking than native simian brilliance. But other animal feats
seemed to imply that animals could retain in their minds a symbolic
representation of the world, information that they could not only recall
but transform and apply to novel circumstances. In one of the earliest
of these experiments, dogs were placed behind a window where they
could see three doors, each with a light bulb above, and were trained to
go to the door indicated by a brief on and off blink of its corresponding light. If the dogs went directly to the indicated door when the window was opened they got a food reward. In itself, no surprises here; this
is a classic learned association. But the experimenters then tried delay-
ing the opening of the window. Up to ten seconds would pass between
the blink of the light and the opening of the window. The dogs' ability
to respond accurately even with a "delayed reaction" suggested that
they were not responding directly to a stimulus, but rather to some re-
tained memory, or mental "representation," *of* the stimulus.

Even more striking were experiments on maze learning in rats. Rats
allowed to explore a new maze even when they were not hungry or

thirsty appeared to retain a "cognitive map" of its layout and where food and water could be found: when hungry or thirsty animals were later let into the same maze they had earlier explored, the hungry animals went straight to the food and the thirsty animals straight to the water. This was termed "latent learning"; even without any direct stimulus, the rats had learned information they could use later.

Neither of these accomplishments may strike us as particularly surprising or brilliant. But both caused genuine problems for the behaviorists' stimulus-response model. There was clearly something going on in the *minds* of the dogs and rats—something we still could not observe directly, something we could only infer, but something nonetheless.

WHAT'S THE DIFFERENCE BETWEEN AN ANTIAIRCRAFT GUN AND A CRICKET?

The behaviorist "taboo" had warned against speculating about what such somethings might be precisely because we had but a single model to draw on: introspection from our own human minds. And such extrapolation from human problem-solving and reasoning abilities had invariably led to embarrassing incorrect conclusions about animals' mental processes, chiefly because human language is both unique and the basis of so much of our problem solving and reasoning. In fact, we scarcely know how to muse about what our minds do without using language ourselves.

Beginning in the 1930s and 1940s another model for thinking about decision making and information processing started to emerge. For millenniums, as the psychologist Howard Gardner has pointed out, our model of the underlying basis of logical operations was the classical syllogism of Aristotle. The salient characteristic of syllogisms is that they use language. But equating all decision making with an ability to manipulate linguistic components had the incidental effect of polarizing the positions of the behaviorists and mentalists. If it is impossible to have reasoning without a humanlike mind, then either you assume that animals do not have any such ability and thus are incapable of making decisions; or you assume they are capable of making decisions and thus have humanlike minds.

Computers for the first time offered a third alternative. As early as the late nineteenth century, mathematicians had formulated a symbolic logic that no longer relied on language at all. By 1936, Alan Turing (inventor of the Turing test of consciousness we encountered in the introduction) proved that, in principle, a machine that performed an almost rudimentary set of logical operations could solve *any* calculation. The so-called Turing machine was a theoretical machine that could move a paper tape, marked out in a series of identical squares, past a scanner. The scanner could tell whether the square it was reading had a mark on it or not, and the machine itself could perform just four operations: it could move the tape one square to the left: move the tape one square to the right; it could place a mark on the square: or erase a mark from the square. These were all binary operations, and Turing's conception of representing data and logical operations as a combination of 1-or-0, on-or-off "bits" (short for "binary digits") became the basis of the modern digital computer.

Meanwhile other researchers, most notably the mathematician Norbert Wiener of M.I.T., began working on the development of mechanical and electrical systems that could be used to steer weapons automatically. These systems could, for example, adjust the rudder and elevators of an airplane to keep it automatically on a preset course. The basic principle of all such devices is feedback. If the compass indicates the plane is drifting to the left of its preset course, the control system turns the rudder to the right until the plane is back on course. I have a toy motorized "mouse" that does much the same thing. It has two electric motors, one for each wheel, and is wired so that it will always follow along a wall to its left. A spring-loaded "whisker" feels for the wall; if it is not pushing against anything, a switch that the whisker is connected to sends current to the right-hand wheel, making the mouse turn left. If the whisker pushes against a wall, it rotates the switch to send the current to the left-hand wheel.

Wiener and others argued that such machines were analogous to the self-regulating processes of the human nervous system and, more controversially, that one could literally speak of such feedback machines as purposefully "striving towards goals."

Whether it makes sense to talk of machines having goals, there is no denying that even simple machines can behave *as if* they have goals. All of these mid-twentieth-century advances in the mathematical theory of computing, and the construction of actual computers, had the effect of redefining Descartes' characterization of animals as mere automatons—for automatons no longer seemed very "mere." They could clearly do a great deal. Indeed, what really began to excite cognitive scientists—what ultimately triggered the so-called cognitive revolution in psychology by the late 1950s and 1960s—was the idea that *human* mental processes, up to and including even consciousness itself, might at last be placed on an explicable, material basis. One key theorem showed that any logical operation that connects any specified combination of input signals to any output could be performed by a finite-sized network of on-off electrical relays—or for that matter a network of neurons. As long as the operation could be specified in unambiguous language, it could be realized by such a "neural network."

Machines programmed with logical operations were not just slavishly carrying out an unvarying sequences of rules—they were not just very fast adding machines—but could be instructed to "have goals" and "make decisions" and alter the course of their operations, on their own, according to the nature of the problem. (Even the most elementary computer programs include such "IF, THEN" instructions. A program that prints out checks to suppliers might include a line that instructs the computer to determine if the amount is greater than $2,500, and if so print "TWO SIGNATURES REQUIRED" on the check. If A, then B.)

There was no reason in principle that a computer could not perform even complicated tasks of symbolic manipulation, such as solving geometry problems or proving mathematical theorems. In 1955, researchers Herbert Simon and Alan Newell worked out a programming language that could be used for writing a theorem-solving program, and showed—by working through one example by hand—how a computer following such a program could actually come up with a proof of a theorem in symbolic logic. "Over Christmas, Alan Newell and I invented a thinking machine," Simon told his class. They subsequently actually ran their program on a digital computer, and showed that the Logic Theorist, as it was dubbed, could indeed produce correct mathematical proofs, and often did so in under a minute. One of the proofs was shorter and more elegant than the previously published, human-produced version.

A final ingredient in the cognitive revolution was the recognition that the physical structure and action of nerve cells permitted the mind, too, to act (at least in principle) as a system of on-off logical switches. A nerve cell fires or it doesn't fire (on-off), and is connected to other nerve cells in such a way that the whole network of such switches was in theory no different from a network of electrical relays. Such networks had already been demonstrated to be capable of performing any calculation that could be precisely formulated by symbolic logic. And symbolic logic had in turn been shown to be capable of representing any syllogism that could be precisely stated in language. The brain was a computer, and a computer could be a brain.

KNOWING WHAT YOU'RE DOING

Here was an entirely new way of thinking about the brain and what it does. At once it seemed to promise an ultimate explanation of the mind—as a literal computer with neurons taking the place of relays or vacuum tubes or silicon chips—and to raise the possibility that a computer could be built that literally duplicates the human mind. Neither claim has precisely held up the way the early enthusiasts expected. Computers have turned out to be very good at some things, but pathetically bad at others. A computer can beat the world chess champion, but cannot equal a baby's (or a sheep's) ability to recognize faces or to walk across a yard without bumping into things.

But the effort to simulate human intelligence with computers has driven home several points of lasting importance. Above all was the key insight of Norbert Wiener's: that it was possible to think usefully and productively about information and information processing totally separately from its physical representation. Whether it was electrical relays or mechanical linkages or nerve cells that did the actual work, information was information and bits were bits. Airplanes used to use wires and pulleys and bell cranks to convert an input from the pedals into a turn of the rudder; new aircraft use an electrical signal altered by the pedal action to activate a hydraulic pump to move the rudder. But in both cases the information comes in, is processed to maintain an appropriate ratio between the amount of pedal movement and the amount of rudder displacement, and an output signal produces the result. We can accurately describe what happens to the information between pedal and rudder without any reference to what piece of hardware intervenes between pedal and rudder. We can treat the information processor as a black box.

So even without any understanding, even without any theory to explain how individual nerve cells in an animal's mind work together to produce a given behavior, it became possible to look at what information was coming in, what behavior was coming out, and what sort of information processing occurred in between. In place of the stimulus-response arc of behaviorism, which imagined a simple connection forged between the nerves that received an incoming stimulus and the nerves that controlled an outgoing response, the cognitive approach conceived of inputs, outputs, and operations. Cognition thus represents an intermediate level of analysis; it is the product of (so far) less than fully explicable nerve actions, but that does not preclude us from studying what the mind does, or developing general principles about how it does it. By designing experiments that force an animal to use information not present in its immediate environment, we can probe the information processes by which an animal maps its territory, or recognizes individuals, or uses numbers, or makes lists. Just as linguists did not need to know anything about how the human mind directs the vocal cords to produce specific sounds in order to study the rules of syntax, neither do students of animal cognition need to know the details of the brain's physical operations to study how it makes certain calculations. Information-processing theory has freed us from the need to draw inferences from the human mind in the study of animal minds. As Howard Gardner points out, cognitive science deals with "symbols, rules, images—the stuff of representation which is found between input and output—and in addition explores the ways in which these representational entities are joined, transformed, or contrasted with one another."

The very failure of computers to reproduce many of the things that human minds do underscored another important point. Computers are very good mimics of the sorts of things that language permits us to do—"planning, problem solving, scientific creativity, and the like . . . tasks performed through conscious and deliberate effort," notes Herbert Roitblat of the University of Hawaii. But paradoxically, "people perform them badly and slowly relative to fluent and apparently automatic sensory and perceptual" tasks. Roitblat goes on to note how many of these automatic, "everyday" tasks, which we do not even think about, actually involved quite sophisticated information processing of a kind that computers, executing discrete sequences of programmed instructions, are horrendously bad at.

That does not mean that the information-processing perspective is the wrong way to understand animal minds. But it does mean that the whole realm of nonverbal information processing that goes on in our minds, and which must be the essence of thinking in all nonhuman animals, is unlikely to be well represented by language or by the languagelike logic of digital computers executing sequential instructions. To turn this point around, however, it also means that to suggest that animals lack the sort of conscious reasoning that language confers uniquely upon humans does not in the least imply that animals are mere automatons. Many sophisticated and unquestionably intelligent feats of calculation, analysis, and decision making of our own take place at a very deep, unconscious level. So it must be for animals.

We are so used to thinking of intelligence as the performance of conscious, creative, linguistically based acts of reasoning that it can take some convincing to persuade people how much genuine intelligence resides in automatic, unthinking layers of their minds—and thus presumably in the minds of animals, too. But if you stop and think, you quickly realize how much you do without thinking at all. If you are like most people, you can find your way down a familiar street with almost no conscious awareness of your behavior at all. You almost never think your way through it; unless you're following directions to a place you've never been, you don't say to yourself, "turn left at the white church, then count the fourth house on the left"; you just turn. In unfamiliar terrain, experiments have shown that humans can unconsciously learn their way around both by dead reckoning—having a sense of how far we have walked in various directions—and by recognizing landmarks. We sometimes articulate the result: "I feel that this is the place we're supposed to turn" or "It seems to me we've walked far enough in this direction; we should head across the field now." But we have not been counting our steps or carrying out a geometric calculation.

Likewise, recognizing a familiar face or a familiar voice normally demands no conscious effort at all—the person's name really does seem to just pop into our heads from somewhere. When we struggle to place a face, though, and consciously try to dredge up a memory, we generally fare quite poorly. The conscious process of logical analysis ("where could I have seen this person before; let me run through the list of people I've met who have red hair and a beard; let me think of the people I've met from my business") is clearly a very different one from the unconscious, automatic, and inaccessible process by which the answer is just *there*. That is probably also the reason that computers tend to be so bad at recognizing faces and objects and things. Dogs come in an extraordinary variety of sizes and shapes and colors, yet we have no trouble recognizing all as dogs. Imagine trying to write an explicit list of rules defining a dog, and distinguishing it from a cat or a raccoon or an opossum. Four legs, two eyes, fur, and a long furry tail doesn't get you very far. (That assumes you have already figured out a way to define what fur and legs and tails are.) That such things might even need defining seems funny to us, but that is proof all the more of how much processing takes place at a level beneath our conscious awareness. If you really start to think about how you would program a computer to recognize fur, the problem is far from trivial. Long furry tails would rule out opossums, but cats and raccoons also have long furry tails, and some dogs do not have furry tails and some do not have tails at all. The list of rules and allowable exceptions to the rules would be enormous and there would probably still be individual cases (Chihuahuas, boxers, English sheepdogs) that the list would not cover. It might almost be easier to define "dog" simply by describing every individual breed of dog and forget about general formulations.

Listening to someone speak and transforming their sounds into words; choosing a path along a puddle-filled street; throwing a ball at a target—these are other things that the brain does very well, that computers do not do well at all, and that when we do them, we do at a completely unconscious level.

MENTAL LAPSES

Particularly intriguing evidence of how much our minds do unconsciously is what happens when they stop doing certain of these tasks. The number and variety of bizarre cognitive disorders that humans with brain damage from injuries or stroke suffer from is extraordinary. We take these unconscious processes for granted precisely because they do not happen in a way that our minds express as language, or even abstract symbols, that we can access or even imagine—we only notice them when they're missing in others. In some disorders, the subject loses the ability to remember names of things, but can still speak in grammatical sentences. In others, words lose their meaning altogether, but the person can take dictation accurately and remember how to spell words. In some cases a person's language ability is purely literal; he can categorize and generalize concrete things but has no ability to hold abstract ideas or grasp metaphorical language. In some cases a person can describe nothing about himself but can remember general facts about the world. In one known case a boy was unable to describe what he did each day, saying he remembered nothing, but could write it down—and then be surprised by his account when it was read to him. In some cases a person literally cannot remember what happened the day before, but can nonetheless learn a new skill, like playing the piano or golf. Each time he plays, he is sure it is the first time he has ever done so, yet each time he gets better. In other cases the opposite happens; a person can remember the fifteen piano lessons he has had, but each time plays as if it were the first time.

Perhaps the oddest of all cognitive disorders brought on by brain damage is the Capgras delusion; victims of this syndrome become convinced that their husband or wife or other close acquaintance is actually an impostor. Yes, this person looks like his wife, and acts like his wife, and even insists that she is his wife, but the victim of the Capgras delusion is absolutely certain that she is not his wife. In a few tragic cases people with this disorder have actually killed the "impostor" that they believe has usurped the place of their loved one.

The neuropsychologist Andrew Young has proposed an intriguing explanation for this syndrome that points to an underlying, unconscious cognitive mechanism for recognizing faces which parallels the conscious process. People who suffer from another bizarre cognitive disorder known as prosopagnosia are unable to consciously recognize familiar faces. When shown an assortment of pictures of celebrities, family members, and random strangers, they are unable to attach a name to any of the faces. Yet, when read a list of possible names, they do show an unconscious physiological reaction when the correct celebrity or family member is named. (This is like the form of "lie detector" test known as the guilty-knowledge test, in which mentioning some concealed fact causes the skin resistance to change.) It seems that at some unconscious level, prosopagnosics actually *do* "know" faces, but for some reason that knowledge cannot be accessed consciously. Young suggests that people who suffer from the Capgras delusion have exactly the opposite condition: consciously they are able to recognize familiar faces, but the unconscious process that continues to operate in prosopagnosics has failed in their case. The disjunction between the conscious recognition and the unconscious nonrecognition is resolved by arriving at the conclusion that although the person certainly looks right, it is not really the person it appears to be.

Other sorts of quirks and logical errors that appear even in normal people are so universal that they, too, strongly imply an unconscious level of cognitive processing that plays a huge part in running the business of our brains. Anyone who has tried to teach statistics knows how difficult a time human beings seem to have with even its most elementary precepts. Even intelligent and well-educated people consistently make the same mistakes.

The most common error is the virtually universal, gut-feeling belief that if you flip a coin and get four heads in a row, the next time it is more likely to come up tails. Go to a casino and you will observe this principle in action all the time; nobody wants to play a slot machine that has just paid off; no one wants to bet on red when red has just come up several times in a row; a losing blackjack player keeps playing because he is "due" to have his luck turn. All of these beliefs directly defy the most elementary fact of statistics. The roulette wheel has no way of knowing that red came up the last four times; each time the chances of red and black coming up are exactly the same. So why do we all make this mistake?

The answer is probably that we are subconsciously making an association between coincident events in the real world. It is after all an observable fact that a series of heads, heads, heads, heads, heads is a rare event. It is an observable fact that slot machines rarely hit a jackpot two times in rapid succession. But underlying statistical laws are not obvious, and are one of the minority of things in the real world where appearances of cause and effect really are deceptive. We are attuned to noting things that happen together because they usually reflect genuine cause and effect. But roulette wheels are built precisely to defy appearances (and the usually reliable circumstances of nature in which things that happen together are causally related), and to produce results that are totally unaffected by what has passed before. The reason that two jackpots in a row are rare is a consequence of multiplying individual probabilities, each of which is unaffected by what has passed before. We look at the end result of the statistical laws operating on a series of individual events and assume that the end result is the cause. But "assume" is actually the wrong word; we in fact sense these things very much at a gut level.

Likewise for the large class of biases in human cognition studied in the 1970s by the psychologists Amos Tversky and Daniel Kahneman. They found that people consistently judged two different situations that were identical from a formal logical or mathematical point of view quite differently depending on their social context. For example, when gas stations started imposing a surcharge of five cents a gallon for using a credit card, customers were angry and unhappy. The gas companies' solution was to raise the price of gas five cents and offer a five cent "discount for cash." Customers were happy again.

Or ask people how they would judge these two situations. Case A: You are on the way to see a play, having purchased a ticket in advance at the theater for $40. On arriving at the theater, you find you have lost the ticket. Do you buy another ticket for $40? Case B: You are on the way to the theater to buy your ticket. On arriving at the theater you discover that $40 has fallen out of your wallet. Do you still buy the ticket? The net result is the same in either case—you have one ticket and you're out $80. Yet most people say they would be much more likely to buy the ticket in case B than in case A.

Tversky and Kahneman suggested that the reason people accept the loss of the cash but not of the ticket is that they keep separate "mental accounts" for the cash and ticket expenditure. But another way of explaining it is that people have an unconscious "cheating detector." We all hate to feel that we've been played for a sucker, that someone is taking advantage of us, and there is much evidence from other psychological studies that people are acutely sensitive to perceptions of fairness. In the case of the gasoline prices, the critical fact is that we resent being charged extra for a service we think should be free. In the case of the theater tickets, what we really may be resenting in case A is that the theater has ended up with $80 while we have only one ticket in return. In case B, there was no such obvious unfairness; the $40 on the street did not wind up in the hands of the people we were trading with.

Again, all of these examples are in the category of gut feelings. Our judgment does not seem to be the product of logical analysis; it wells up from somewhere beneath our consciousness. There are wheels spinning all the time, making judgments about the world, and about the behavior of our fellow humans in particular. We have all probably had the experience of knowing that someone is lying to us without our even knowing exactly what it was about their behavior that made us feel that way.

MENTAL MACHINERY

Some detailed and well-controlled psychological studies of human perceptions and mental reactions have turned up further evidence of unconscious cognitive processes that may be examples of the sort of unconcious and nonverbal thought processes that take place in animal minds. One of the earliest and most famous examples was that noted by the psychologist George Miller in his 1956 essay titled, "The Magical Number Seven, Plus or Minus Two." Miller observed that in study after study of human psychology, people seemed to be able to remember and juggle and compare up to seven different items: beyond that, their ability fell apart. It didn't matter whether it was numerical digits or lists of words or items to compare, seven seemed to be the magic number. Miller's rather whimsical opening statement read, "My problem is that I have been persecuted by an integer. For seven years this number has followed me around, has intruded in my most private data, and has assaulted me from the pages of our most public journals."

When we try to remember things that contain more than seven items, we often resort to "chunking" several items together. It is interesting to look around and see just how many examples of the rule of seven there are in the world, and how we try to get around this apparent limitation. The dashes in phone numbers are one way we try to "chunk"; we tend to remember area codes as a single familiar unit. Mnemonic devices and acronyms are another form of chunking. The musical scale of twelve notes is reduced to seven letter names, the other notes filled in by calling them sharps and flats. It may just be a matter of what we are used to, but a notation system with twelve names of notes (letters A through L, say?) would seem awfully confusing.

The point about all this seven business is that it implies something very definite about the underlying cognitive architecture of our brains.

Information is processed in a certain way, beneath our ken, but with very real consequences.

Another class of studies that is strongly suggestive of unconscious processing involves mentally rotating objects. A subject is shown an image—a capital letter F, say—along with either a rotated F, or a rotated mirror-image of an F:

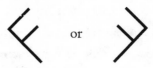

The subject is then asked whether the object matches the original. People generally get the answer in less than a second, but the remarkable finding is that the exact time it takes a person to answer the question is directly proportional to the amount of the rotation. If you rotate it 40 degrees, it takes a person twice as long as if you rotate it 20 degrees. In effect, people seem to be rotating the image in their minds. As we shall see in the next chapter, this experiment has been repeated with pigeons and apes; it is the perfect sort of cognitive experiment because it taps directly into a computational process of the mind. And the quantitatively consistent results offer a high degree of confidence that there is something real and fundamental going on there; they also even make it possible to speculate about what sort of "algorithm" the mind is programmed with to perform such calculations.

The idea that other such elementary cognitive operations are performed by specific "subroutines" in the brain has received further confirmation from studies showing that different sorts of elementary operations require different amounts of time and appear to take place in different parts of the brain, as well. For example, subjects were first shown two letters and asked to answer as quickly as they could whether they matched (A and A, for example). On average it took them about 0.55 seconds to answer. Next they were shown an upper case letter and a lower case letter (A and a) and asked if they were the same letter; this took 0.62 seconds. Finally they were shown two letters (a and e) and asked if they were both vowels or both consonants; it took 0.7 seconds to detect matching vowels, 0.9 seconds to detect matching consonants, and

0.8 seconds to detect mismatched pairs. Strikingly similar results were obtained when pairs of plant and animal names were used (MOOSE and MOOSE; MOOSE and moose; moose and rabbit). In each case the first category requires only a physical comparison of the two items. The second requires matching identical names that have different physical representations, and the third requires applying a category rule. (That there are more consonants than vowels may explain why it takes people less time to recognize pairs of vowels than pairs of consonants.)

Even more interesting, brain scans have shown that different regions of the brain are automatically activated during different unconscious tasks associated with words. Positron emission tomography (PET) scans, which measure increased blood flow through the brain, show that certain regions of the brain are particularly active when subjects are asked to look passively at a screen on which a series of nouns appear. Additional, and quite localized regions of the brain become active when the subjects are then shown a series of nouns and are asked to perform some "semantic" operation—such as to say aloud a use for the named object or to note which of the words on the list are dangerous animals. Brain scans similarly reveal that different areas of the brain are activated when subjects are shown faces in various positions within a square box and asked either to identify the face or state its location in the box. "What" and "where" seem to be processed separately. There appear to be many other unconscious ways in which our brains encode and categorize information. For example, if subjects are shown pictures of animals, and asked to say the names of them silently to themselves, one part of the brain is activated; show pictures of tools, and another part of the brain switches on—this latter the same part of the brain that is activated when people are asked to imagine the hand motions required to perform a manual task.

Much of this kind of evidence suggests that the brain is using algorithms very different from the kind a digital computer, programmed with sequential instructions, would carry out. A mental rotation is a sort of visual analogy, not a series of symbolic-logical operations at all. The mind here is employing a "representation" and "operation" that is neither linguistic nor symbolic, but perhaps pictorial—or perhaps something even deeper. The question of *what* a mental representation *is* lies at the heart of the mystery of animal thinking and consciousness.

WHAT ROBOTS CAN'T DO

As we have seen, digital computers proved phenomenally bad at performing many of the mental tasks that brains do with such automatic ease that we are not even aware of them. The simplistic claim that a general-purpose computer could be programmed with a set of instructions to mimic the human mind's many feats is in bad odor these days, both for practical and philosophical reasons. It became apparent that some sort of specialized framework is probably critical to such tasks as visual perception. A list of symbolic-logic operations for recognizing a dog—or even a vertical line—is tedious in the extreme. Even when it works it seems to take forever compared to the time it takes a person to perform the same task—even though an individual computer memory bit can work much faster than a neuron. By contrast, wiring nerve cells (or computer bits) together in the right way to begin with can produce a specialized "neural network" that automatically makes such specialized calculations as recognizing a vertical line. The evidence is mounting that the unconscious parts of the mental computer are not some general-purpose computing machine that executes symbolic logic operations in sequence, but rather a set of specialized machines built around an architecture of "distributed" processing. The answer to a calculation isn't in one bit somewhere at the end of a sequence of logical operations executed in series; it is the state of the whole network that determines the answer.

The other objection to the sequential, logical processor model was that many of the programs that were claimed to demonstrate artificial intelligence were really more parlor tricks than genuine mimics of the human mind. The most famous example of pseudo-intelligence produced by the standard symbolic-logic approach was the program called ELIZA, created by M.I.T. computer scientist (and artificial-intelligence skeptic) Joseph Weizenbaum. ELIZA was programmed with a limited number of fixed "IF THEN" rules to generate a superficially quite convincing facsimile of a psychiatrist conducting a therapy session. A human "patient" would type in a sentence, and ELIZA would respond with seemingly apt questions to further probe the patient's troubles. A typical dialogue ran like this:

> Human: Men are all alike.
>
> ELIZA: IN WHAT WAY?
>
> Human: They're always bugging us about something or other.
>
> ELIZA: CAN YOU THINK OF A SPECIFIC EXAMPLE?
>
> Well, my boyfriend made me come here.
>
> YOUR BOYFRIEND MADE YOU COME HERE.
>
> He says I'm depressed much of the time.
>
> I AM SORRY TO HEAR THAT YOU ARE DEPRESSED.
>
> It's true. I am unhappy.
>
> DO YOU THINK COMING HERE WILL HELP YOU NOT TO BE UNHAPPY?
>
> I need some help, that much seems certain.
>
> WHAT WOULD IT MEAN TO YOU IF YOU GOT SOME HELP?
>
> Perhaps I could learn to get along with my mother.
>
> TELL ME MORE ABOUT YOUR FAMILY.

The somewhat scary thing is how many people, when first exposed to this program, started pouring out intimate details of their personal lives. Even many mental-health professionals took seriously the idea that ELIZA was the forerunner of machines that could truly perform psychotherapy.

The program was easily fooled itself, however, for it was supplied with a relatively small number of rote rules. Certain key words would trigger apposite statements (any occurrence of *brother, sister, mother, father* would lead to the statement, "Tell me more about your family," for instance). Sentences of certain forms would be transformed to questions by applying rules such as turn "my" into "your." But make a syntactically tricky remark and the game was up. The statement "I am hardly ever happy at all" might elicit the tongue-twisting question, "DO YOU THINK COMING HERE WILL HELP YOU NOT TO BE HARDLY EVER HAPPY AT ALL?"

Daniel Dennett describes a similar parlor trick that he and colleagues at the Artificial Intelligence Lab at M.I.T. cooked up. Their project has the ultimate goal of producing a "conscious" robot. At his current state of development, though, "Cog," as he is nicknamed, is nothing of the sort. Nonetheless, he has been tricked up with some disturbingly human attributes. "Cog cannot yet see or hear or feel at all," Dennett says, "but its bodily parts can already move in unnervingly humanoid ways. Its eyes are tiny video cameras, which *saccade*—dart—to focus on any person who enters the room and then track that person as he or she moves. Being tracked in this way is an oddly unsettling experience, even for those in the know. . . . Cog's arms, unlike those of standard robots both real and cinematic, move swiftly and flexibly, like your arms; when you press on Cog's extended arm, it responds with an uncannily humanoid resistance that makes you want to exclaim, in stock horror-movie fashion, 'It's alive! It's alive!' It isn't, but the intuition to the contrary is potent."

The point is that some simulations of behavior and intelligence are not real models at all of what a living being does; they are missing fundamental ingredients and thus offer no insight at all into how a living brain is performing the same task. But they capture just enough of the real thing to be able to fool us, at least for a while.

WHAT ROBOTS CAN DO

So we have to be always on guard when we think we have come up with a computer model of live intelligence, for it may be simulating far less than we think. All of these cautions notwithstanding, computer simulations are proving a powerfully insightful tool for testing ideas we may have about how an animal performs a particular cognitive task. One consistent story they have told is that even complex computational tasks can take surprisingly few neurons (or the computer equivalent), provided those neurons are wired together the right way. Computer simulations using fewer than a hundred neurons, with fewer than two hundred connections between them, have demonstrated some of the adaptive capabilities of real insects. And these adaptive capabilities go well beyond simple stimulus-response associations. Recall Roy Caldwell's observation: It doesn't take a lot of neurons to be deceptive.

These computer models of animal cognition and behavior have been dubbed "animats." Some exist only as computer programs run on a screen; others have actually been built as moving, sensing robots that interact with their environment. One of the simplest is a computer model of the bacterium *E. coli*, which lives in the human gut. *E. coli* have an extremely primitive method of navigation. They can propel themselves in a straight line in the direction they already happen to be heading, or they can go into a tumble that changes their direction—but the direction they end up in is entirely random. They have no way to make controlled left or right turns. Yet, as an exceedingly simple computer model showed, these two crude methods of locomotion, combined with a brief memory, are all *E. coli* needs to find its way to a chemical attractant (a food source) or away from a repellent (a toxin). The model assumes that the *E. coli* follows this simple algorithm in moving toward an attractant: It senses the strength of the attractant at one spot, moves in a straight line, senses the strength of the attractant again, and compares its two readings. If the signal is getting stronger, it keeps moving in a straight line. If the signal is getting weaker or not changing, the bacterium tumbles randomly and then repeats the process. What the simulation showed is that this is literally all it takes to move reliably toward the source of food. Humans asked to perform the same task—to guide a dot to a goal on a computer screen by tapping a key that produces random changes in direction—produced tracks indistinguishable from the simulated *E. coli*, or real *E. coli*, as well.

The important lesson is that the model fully reproduced *E. coli* behavior in zeroing in on a goal even though it had no "map" of its environment, no plan for its trajectory, no "knowledge" of concentrations, or odor plumes, or gradients—or even directions, for that matter. "It lives in a world of briefly remembered one-dimensional signals," says W. Thomas Bourbon of the University of Texas, who built the computer *E. coli*. Bourbon says that many of his fellow scientists who were told of these results simply refused to believe that such a simple and crude algorithm could allow a life-form to successfully orient itself in its environment; encountering such resistance was what prompted him to have people try "being" *E. coli* for themelves to see that his algorithm really does work to navigate toward a goal.

The *E. coli* animat does so much with so little because it takes advantage of information contained in its environment. The same principle appears time and again in other animats. The basic idea is to avoid having to solve a problem from scratch, and instead find ways to combine the simplest possible sensory inputs in ways that produce the right behavioral response. For example, suppose you (or evolution) were trying to "design" an animat (or an animal) that would always move towards a source of light. One way to do that would be to equip your creature with a rotating eye that could scan the horizon, repeatedly measuring and recording the intensity of the light; after it had completed a full circuit it would have a "map" of light intensities in all directions, and could then evaluate the data and determine which direction corresponds to the greatest intensity. Then it could calculate the difference between that direction and the direction the creature itself is currently facing, and give the appropriate commands to its feet to turn the correct amount and start moving forward.

Aside from the computational complexity of this approach, it has other disadvantages. For one, the light source might move. Also, it's very easy for small errors to creep in. The ground might be uneven, causing the creature to shift off course as it starts heading forward on its assigned compass direction. Even transferring the calculated course to the feet in the first place can introduce errors if the eye compass and the feet compass are not perfectly aligned. Imagine steering a sailboat toward an object that is in view by first measuring the angle between the object and the ship's bowline, then using the ship's compass as your guide to turn the boat that many degrees left or right from your current heading.

A far simpler—and foolproof—method is to design the creature so that course corrections are made automatically, through a feedback process that is self-correcting. Instead of a rotating eye, use two eyes, and simply let them compete: As the intensity of light striking the left eye increases, so does the signal that eye sends to the right feet. Likewise the right eye sends a signal to the left feet. So if the light is off to the right, the left feet will move faster than the right, causing the creature to turn to the right. When the light is dead ahead, the signal from both eyes will be the same, and it will move straight ahead.

In place of calculations and mapping, we have a system that takes advantage of a salient feature of the environment and a simple feedback system that puts that information to direct use. If the light moves, the system automatically adjusts to follow it. If it gets bumped off course, it automatically gets back on course. You don't have to know precisely where a light is to know that when your right eye is receiving more light than your left, it's somewhere to the right. Nor do you have to have any knowledge of maps or paths or spatial relationships or know that the light you are tracking is coming from a particular point in space or that it may be in motion. All of that is folded into the design of the creature's sensors and feedback loops. To refer back to the sailboat analogy: You don't need to be able to read a compass or measure angles with a sextant or look at a map or calculate a course in order to steer the boat to an object that is in view: If it's to the left of the bow, you turn left. If it's to the right, you turn right. That's it.

Animats become particularly interesting when the interactions between the various feedback loops they have been equipped with reproduce complex, and realistic, behaviors that are greater than the sum of their preprogrammed parts. An animat frog, for example, was programmed with three simple rules. If a prey animal is straight ahead and within range, the frog's sensors generate a signal to snap with its mouth to grab it. If it is out of range and straight ahead, a signal tells the frog to hop forward. If it is to the side, a signal is sent to turn toward it, while inhibiting a hop or snap at the same time. The strength of each of these three signals is proportional to the strength of the sensory input. A fly very far ahead will stimulate a long hop. A fly very far to one side or another will stimulate a big turn while strongly inhibiting a hop or snap.

Because all three of these signal channels run simultaneously, competing and summing with one another, their net combination can produce novel and complex behaviors. For example, if a fly is very close to the frog and just a bit to its right, the turning signal is tiny, and only slightly inhibits the very strong snap signal; as a result, the frog snaps while turning to the right. If the fly is just barely out of snapping range, the hop signal is weak and the frog takes only a short hop; that hop is then immediately followed by a snap as the frog enters snapping range. The remarkable thing about all this is that none of these "creative" and aptly competent combinations of behaviors are controlled by a "top-down" command from some central controlling brain that assesses the total situation and plans a course of action. Rather they are the product of a distributed process of competition and cooperation between simple elements that are permitted to combine freely with one another.

There is no guarantee, of course, that this is the way a real frog is "wired." But animats certainly demonstrate that simple control loops that interact with the environment can produce appropriate, complex, and even creative behaviors. Animats are a realization in hardware of Morgan's canon—they show how psychological processes far simpler than we had thought possible can account for observed behavior. That they often show capabilities that even their inventors have a hard time accounting for, given the simplicity of their component parts, is itself an exciting and rather stunning discovery. Perhaps the linear, linguistically based conscious mind of humans finds it hard to intuitively grasp just how much a distributed, nonlinguistic control system can accomplish. Griffin's equation of decision making with conscious, centrally directed thought seems more questionable than ever.

WIRING A CRICKET

When the wiring of an animat is modeled closely on what we actually know about the neurophysiology of the animal it is supposed to model, the demonstrations become even more suggestive. They are a way of testing possible theories of how an animal performs its cognitive tasks, within the actual constraints that physiology imposes. By the way: Even if we know everything there is to be known about the physiology of the brain and the sensory organs, we can never understand what is going on in there without developing and testing theories at the cognitive level. Again, this is the problem of understanding what a computer program is doing by studying the computer wiring diagram and measuring voltages. What we really want to understand is the logical processes, the mathematical algorithms, that connect input and output. Looking at the neurons is not enough; we need to understand the concepts that underlie the wiring of the neurons.

But the more closely our cognitive theories are based on physiological reality, the more confident we can be that we are on the right track. A revealing case study comes from efforts to understand how a female cricket locates a mate. The females show a number of remarkable abilities. A female cricket can pick out the song of a male of its own species from the very similar songs of other species. Experiments have shown that females respond most strongly to chirps that are repeated in a very precise rhythmic pattern that is unique to each species. Females are also able to move accurately toward the loudest and closest male of their species even when many other competing males are filling the air with their own serenades.

It sounds like a very complex cognitive task. The female has to sort through a discordant chorus of competing sounds, pick out the ones that have the right rhythm, and then additionally pick out the ones of those that are the loudest, and then to top it all off locate where that one sound among many is coming from. But researcher Barbara Webb found that by incorporating what is known about the cricket's anatomy and neurophysiology—its ears and auditory neurons in particular—a computationally very simple animat model could be made that almost precisely imitates the behavior of a female cricket homing in on a male.

Cricket ears themselves have a number of peculiar properties. For one, sound arrives at each eardrum via two different routes—directly through the air, and indirectly via an internal tube that opens out to the air at the center of the cricket's body. The sound waves that arrive at each ear are thus a blend of two waves that have traveled different distances. Imagine two series of ocean waves meeting from slightly different directions. If the peaks and crests line up exactly, you get one series of even stronger waves. But if they are a little bit out of phase with one another, they will interfere and cancel. The length of the cricket's ear tubes are just right to have two acoustical properties. First, they cause a phase shift: on the ear closest to a sound source the two incoming waves reinforce one another, while on the off side they cancel. Second, this phase shift only happens with sounds of one pitch—a pitch that just matches the pitch of the male's song. So other sounds appear to the cricket to be equally loud in both ears, no matter what direction they are coming from. Only a sound of the right pitch produces a different apparent loudness in the left versus right ear that depends on its direction.

So a good half of the cricket's computational problem—telling sounds of the right pitch from the wrong pitch, and telling where those right sounds are coming from—is handled not by the brain at all, but rather is taken care of by the physical design of the ear itself.

Neurophysiological studies have provided further clues to the architecture of the cricket's auditory system. Each ear contains about fifty neurons that pick up the ear's vibrations. These fifty neurons then feed into a single "interneuron." Direct measurements with electrodes have shown that the interneuron begins firing once the total input from the fifty auditory neurons reaches a certain threshold. The louder the sound, the faster that threshold is reached, and the faster the interneuron continues firing as well. This research with actual crickets has shown that whenever the cricket is moving in response to a sound, it turns toward the side whose interneuron is responding most strongly.

Armed with this information, plus Lego blocks, two motorized wheels, microphone ears, and a 68000 microprocessor, Webb then proceeded to build a working cricket robot. And she hit on a cleverly simple algorithm that applied what was known about real crickets and their nerve cells. The algorithm not only detects which side has the strongest signal, but also automatically filters out signals that don't have the correct rhythm. Webb assumed that it wasn't the repetition rate of the interneurons that controlled the steering decision, but rather which one of the two fired first. The first to fire is the loudest, just as the most rapidly firing one is. But focusing on which one is first gives additional information that is not captured by the repetition rate. Imagine what happens, for example, if instead of a rhythmic chirping the incoming signal were a continuous tone. The which-side-repeats-the-most cricket would simply steer toward it. But the which-side-is-first cricket would lose its way almost immediately; it is only at the beginning of the signal that it can compare which side fires first. A continuous sound, or one that repeats so rapidly that the cricket's neurons have no time to recover between chirps, would be filtered out. Likewise, a sound that repeats too slowly might not give enough steering information to allow the cricket to reach the male without a considerable amount of wandering off course.

Webb's cricket robot was programmed with only a hundred lines of code. Some of those instructions were required just to simulate effects like the neuron recovery period that are actually part of the physical hardware in a real cricket, so the actual complexity of the algorithm itself was even simpler than it appears. When placed at one end of a ten-by-twelve-foot "arena" with a speaker at the other end playing chirps, Webb's robot performed very much like a real cricket. When she increased the chirp rate, the robot made little attempt to steer at all, just as you would expect since the sounds in effect blurred together as a continuous tone which offered almost no "which side is first" information. When she reduced the chirp rate, the robot headed in the general direction of the speaker, but followed an arcing path that usually failed to make it all the way there. Real crickets show exactly the same behaviors.

The robot showed two other behaviors that were not deliberately programmed into it but which closely reproduced real cricket behaviors. Webb was curious what would happen if the cricket had to "choose" between chirps coming from two speakers. She knew it was not actually capable of sorting out or distinguishing between multiple sounds; the robot had no algorithm to accomplish this task and it seemed likely it would simply become "confused." Yet, in trial after trial, Webb found, the robot "seemed to have no problem making up its mind (so to speak) and went almost directly to one speaker or the other." The interaction of the robot's simple mechanisms with the environment was able to produce a far more complex behavior.

The other experiment Webb tried was to see what happens when a series of three or so chirps was separated by a period of silence, which is what real male crickets do. Again, in her design and programming of the robot she had completely ignored this complication. But when presented with these more realistic chirp patterns, her simple robot nonetheless did quite well. In fact, it usually arrived at the speaker faster than it did without the silent periods. Apparently the most efficient strategy for a maneuvering cricket is not to make repeated little course adjustments, but rather to get enough information to set its course, follow that course in a straight line for a while in silence, and then get another course update. The robot would sometimes miss or overshoot the speaker a bit under these circumstances, but even when it had to turn around, it got there faster than when it made a continuous series of finicky course corrections every step of the way. So it turns out that no additional cognitive processing mechanisms are required to explain this at all—it is simply an additional consequence of the interaction of simple cognitive algorithms with the environment. From an evolutionary point of view, this raises an important cautionary note. Much of the complexity of observed behavior, especially communicative behavior, reflects just one species, or one sex of a species, exploiting the wiring of another organism. Female crickets weren't necessarily built to respond to a series of chirps with silent intervals between. But males that added that silent interval were more likely to have a mate come running faster, so over the course of evolution males that had chirped that way were more likely to mate and pass on this quirk to their male offspring. Trying to search for how the female cricket's brain is able to respond to these special chirps would have been to start at the wrong end—because that response was probably just a coincidence in the first place.

WHAT DOES A CAT SEE IN ITS MIND?

The scientific study of animal cognition is based on the notion that some things animals do depend on their having mental representations of the world that they can manipulate independently of stimulus-response behavior. Certainly our real everyday curiosity about animal minds is very much wrapped up in our wondering what animals see and understand and feel in their minds. But what might these representations look like? What does an animal see and think about and sense and feel?

The conservative answer is: nothing at all. That does not mean that animals lack sentience, or that they really are nothing but dumb machines. What it means is simply that there may be no meaningful way to describe what an animal's representations of the world are. We know that animals lack a linguistic representation of the world. But nonlinguistic representations may take a form that is literally indescribable—neither visual, nor symbolic, nor anything. Consider the animat models: The animat fly-finding frog does in a sense have a representation of space, but it is a representation divided up into the control loops for three different behaviors and the relative strengths of the signals in those loops. The calculation or transformation the frog is applying to the sensory inputs it receives from its environment is built into the very nature of its distributed wiring. As one recent scientific paper noted, "When we say that an animal perceives an intruder, we do not mean that the animal entertains a mental sentence like 'There is an intruder in front of me'; rather we mean that the animal is in a neural state related to the world in a way that an observer can describe by the reported sentence." Webb's robot cricket can find a mate without any visual image of a mate, any auditory image of a male's mating song, any stored map of its terrain, or any sense of direction. We wouldn't be wrong to say that its auditory nerves are wired in such a way that there *is* some "representation" of the right song stored in there, but that representation exists in a functional form that does not lend itself to any images or symbols our imaginations can conjure up. There is no pictorial or symbolic equivalent. We may think up analogies—like lists, or prototypes, or pictures, or sequences—but they bear no more relation to the reality than do the familiar images we use in modern physics to describe subatomic phenomena. "Waves" and "particles" are things we can picture; they are images drawn from our everyday experience; but that's all they are when we are talking about quantum mechanics. Light is not really made up of little particles, nor is it a wave; light in fact is only completely described mathematically, an abstraction that corresponds to no physical picture at all.

Studies in both animals and humans strongly suggest that even literal visual images are not stored in our minds as images. Early studies of the activity of nerve cells in the visual cortex of cats revealed that a substantial degree of visual processing, and categorizing of different types of images, is automatic, wired into the nerves. Nerve cells in the first layer respond only to the presence or absence of light at a particular spot in the visual field. But these nerves feed into a second layer of "complex cells" that respond to vertical or horizontal lines. These in turn feed into "hypercomplex cells" that are activated by certain distinct shapes, such as a corner. Further research found additional cells that respond to still more complex shapes. Frogs have "bug detector" cells that are activated by blob shapes in the visual field. Some cells in monkeys even respond to images in the shape of a monkey's hand or a monkey's face; some appear to be sensitive to a *particular* monkey's face. A firing neuron is not a word or a picture or a symbol: it's just a firing neuron. Does a frog have a "mental picture" of a bug? Or is that mental representation really just a function of its neural wiring? It is clear from both animat models and from research on actual animals that it is not necessary to have knowledge of or beliefs about or visual models of bugs in order to be able to recognize one.

The idea that mental images are in truth a figment of the imagination has been bolstered by research in monkeys showing that single visual images are actually stored in several different locations in the brain. Thus they are not stored as pictures at all, but as disintegrated representations of pictures that again defy description in terms we can understand. Neurophysiological experiments in monkeys have shown that cells in the temporal lobe of the brain are sensitive to an object's shape and color, while cells in the parietal lobe react to an object's location. Destroying the temporal lobes renders monkeys incapable of discriminating patterns, but leaves them with the ability to distinguish locations; destroying the parietal lobes has exactly the opposite effect. Perhaps most interesting of all is research in humans which shows that when subjects are asked to consciously recall visual images, the picture is built up one part at a time. Subjects were shown block letters drawn on a five-by-four grid. They were then shown a blank grid and asked if the block letter would cover one or two of the grid squares marked with an X:

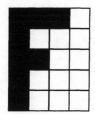 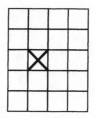

The time it took subjects to answer was directly proportional to the number of line segments in the letter. (An "E" for example has four segments, versus three for an "F.") More time was required when the X fell on the particular segment of the letter that people typically draw last when they are writing a block capital.

Psychological studies have found that human subjects are notoriously unreliable in describing the symbolic or pictorial nature of nonlinguistic thoughts. Accounts differ dramatically from one person to another; some people report their nonlinguistic thoughts taking on rather vivid symbolic forms, others report having no symbols at all. But when these accounts are compared to actual measurements of what the brain is doing, inconsistencies always appear. For example, most people report that when they actively recall a visual image of a letter or other simple object, the image "pops into their head" all at once. But the experiments described above showed that this is not the case at all; the parts of the object "pop into the head" one bit at a time—because they have been stored by the brain as a collection of individual parts plus (in a separate area of the brain) the instructions on how these parts spatially fit together.

What do these separate stored bits of information "look" like? Again, the only true answer is "nothing at all." Most cognitive scientists take the term "representation" as a conceptual notion; their aim is to determine what the calculating mind *does*, and stay away from the unknowable terrain of what an animal subjectively experiences as its mind does it.

However, in some of the examples we will encounter in the following chapters, neurophysiological studies are beginning to explain the underlying nature of the representations and operations animals perform to make maps, recognize objects and individuals, learn lists, and so on. In this way we are taking the only possible steps toward a real understanding of animal consciousness.

4

USING THE OLD NOGGIN

◂◂ A FABLE, WITH A MORAL. OLD B. F. HAD BEEN DRINKING again, and, by some odd quirk of nature, the more he drank, the more his mind would become the perfect exemplar of the behaviorist paradigm. In this condition, everything was stimulus-response, and his friends (cognitive scientists all) began to worry that B. F. should not be permitted to drive in this state. Having to learn by trial and error what the red, yellow, green, and green arrows of a traffic light indicate, they realized, would alone exact a terrible toll on B. F.'s car, not to mention his physical well-being. They briefly considered fighting fire with fire and using a behaviorist paradigm to deal with the problem—a mild shock whenever B. F. touched the ignition key. But they finally hit on what seemed the perfect solution: a special ignition lock that could only be opened by someone whose mind was capable of forming, and transforming, mental representations. They installed a screen that would flash a series of two letters, one at a time, then ask if the letters that had been flashed were A and B. To start the car, B. F. would have to press the correct answer, YES or NO. He would obviously need to maintain a mental representation of both the correct sequence (A and B) and whatever letters had actually appeared on the screen, something no stimulus-response organism could accomplish.

Only a few days later B. F.'s car was wrecked again. He explained that the lock really had required nothing more than stimulus-response behavior to defeat. He hadn't analyzed or compared sequences of letters at all. Through trial and error he had hit on this solution: If the first letter shown on the screen was A, he would stand on one leg. If the second letter was anything but B, he would put his leg back down. When the YES or NO question appeared, he would push YES if he was standing on one leg, otherwise NO. Then he would start the car and drive off. He reported he was still working out what to do when a traffic light showed a left arrow.

Refusing to admit defeat, his cognitivist friends tried again. They added alphabetic keys to the lock mechanism and taught the sober and cognitive B. F. to type in A, B, and C in sequence; when he typed all three letters onto the screen in the right order, the lock would open. This time it took a week, but again, there was the drunken B. F. driving. Again, he explained that he had learned to produce the correct sequence through pure trial-and-error behavioral conditioning. He had learned three simple associations: when the test starts, push A; when A appears on the screen, push B; when AB appears on the screen, push C. Again, he hadn't needed a complete mental representation of the sequence at all.

But finally the cognitivists hit on the solution. This time they taught B. F. a five-letter password, ABCDE. The screen would display a two-letter combination of nonadjacent pairs of letters—AC, CA, AD, DA, AE, EA, BD, DB, BE, EB, CE, or EC—and B. F. would have to key in the two letters in the left-to-right order in which they appeared in the actual password (AC, AD, AE, BD, BE, or CE). As long as they changed the password frequently, this was a task that only the sober, and cognitively functioning, B. F. could manage—he had to keep a representation of the entire list in his mind in order to figure out in which order nonadjacent letters occurred.

B. F. conceded the errors of his ways and swore off drink—and with it, the greater evil of dogmatic behaviorism.

FORCING THE ISSUE

The moral of the story is one that any animal cognitive psychologist would recognize at once. Trying to devise experiments that force an animal to activate an indubitably cognitive mental process is not easy.

In the animat experiments, researchers try to guess at what mental algorithms an organism might be using, then build those algorithms into a robot and feed it the inputs that a real animal receives from its world. If the robot behaves the same way as the real animal it is modeled on does, then we have at least indirect evidence that we have guessed correctly. The beauty of that approach is that the robot's brain is transparent: We devise its cognitive architecture, and we can manipulate it at will.

Real animals are far less cooperative. The only things we can control are the inputs, and the only things we can observe are the outputs, and though the mental process that seems to connect input X to output Y may seem perfectly obvious and straightforward, we have to scrupulously and arduously rule out all plausible alternatives before we jump to a conclusion. The problem that bedevils all efforts to infer cognitive processes in animals is that many mental feats that look to us like "thinking" of a sort that involves performing calculations on or transforming mental representations can be explained as simple learned associations—or even, as we have seen, the unthinking intelligence of evolution.

Some of the first well-controlled studies of animal intelligence looked at whether animals could make a simple discrimination between two stimuli to earn a food reward. Rats were rewarded for turning left in a Y-maze, for example, or pigeons were rewarded for pecking a blue key rather than a red key, or dogs were rewarded when they pressed a lever after one tone was sounded but not another tone. Their ability to master these tasks was of course completely consistent with the simple stimulus-response, learned association model. And in learning such tasks a goldfish does as well as a chimpanzee—corroborating the notion that simple learned associations are likely at work. No cognitive explanations need apply.

But these experiments led to other pioneering experiments that are probably the simplest demonstrations of higher level cognitive processing. One thing that experimenters have repeatedly found is that birds and mammals "learn to learn." It may take dozens of tries to train a horse to push its nose against one of two visual patterns to obtain a reward. But after being trained on several pairs of patterns, the horse seems to catch on to the rules of the game, and it starts learning new pairs much faster. In other words, the horse has not only learned to pick the squiggled lines over the straight lines; it has learned that whenever two visual images are presented, choosing the one that resulted in a reward one time will result in a reward again. Experienced chimpanzees have even been able to master a new discrimination problem in a single trial. Such an ability to generalize clearly implies that something more than simple associative learning is taking place. (The ability of chimpanzees to apply past experience with a single try may be the basis for anecdotal accounts of seemingly "insightful" problem solving by chimps faced with entirely novel situations. Particularly in the case of animals observed in the wild, it is impossible to know what past experience the animals may be drawing on. The ability to apply relational concepts is of course not a trivial cognitive feat—but neither is it a case of reasoning through insight.)

Perhaps even more telling is that primates previously trained with "reversal" tasks generally do better on new discrimination tasks than those who have not had this experience. Animals that master reversal problems (in which the previously correct answer is switched with the previously incorrect answer) appear to have learned an even more general rule: repeat a previously successful choice, but *immediately* switch to the other choice if the first is unsuccessful. In effect they have learned not just that there is food under one or the other of the boxes, but that there is a *rule* that is maintained until a new rule takes its place.

Testing whether an animal can transfer experience from one problem to another is a fascinatingly simple way to tap into its ability to form nonlinguistic concepts. For example, in a test that is so common that it is generally known to psychologists by an acronym—MTS, or match-to-sample—an animal is shown a photograph or an object such as a geometric shape; the animal then has to pick out from an array of objects the one that matches the sample. When chimpanzees are trained to perform this task with one set of objects and then presented with a completely new set of objects to match, they start getting right answers much faster than they did the first time around. Some chimps are able to transfer what they learned from the first set within just a few trials with a new set. In other words, they have apparently learned more than just a series of associations (pick the lamp when you've been shown the lamp, pick the red bug when you've been shown the red bug, and so on and so on); they have rather learned to apply an abstract concept of "sameness."

This "sameness" test can be taken to a further level of abstraction when the MTS test itself is based on a conceptual relation, rather than absolute identity, between matching items. For example, the animals might be trained to match objects that are not identical but which share an attribute such as color or general shape. Or the task may be to pick the one object of a set which is different from the rest (an "oddity" test). Again, chimpanzees and some monkeys do quite well at applying an abstract notion of relationships to new situations. An ape first taught to match objects of the same color can master a test whose goal is to match objects of the same shape in significantly fewer trials than those who lack this previous experience.

In tackling these kinds of problems, species differences do begin to appear. Pigeons for example take far longer than monkeys do to learn to match objects by color or shape in the first place—pigeons need literally thousands of trials before they can perform consistently in making such matches. But more significant than their slow learning curve is the fact that they seem unable to transfer the sameness concept to novel tasks. While they are able to apply the color rule to novel objects (they do well at matching a red square to a red triangle after having been trained to match red circles to red blobs), they require essentially the same number of trials to learn each *new* sameness relation (such as matching objects of the same shape). In other words they have to start from scratch each time: They can learn a single sameness rule, but cannot transfer the concept of sameness.

Similarly, pigeons, rats, cats, monkeys, and apes can all learn a conditional task—for example, pick the triangle rather than the square if they're red, but pick the square if they're white. But only monkeys and apes improve when presented a series of novel conditional problems; the rats, cats, and pigeons have to learn each new one from scratch.

A cautionary note about all of these experiments: As we saw in chapter 1, some of the differences between species may be the result not of cognitive differences but rather of sensory biases built into the very nature of the test, and the final word is not yet in. Rats generally do quite poorly on visual discrimination tasks. They also fail to improve over a series of visual tasks. An obvious conclusion would be that they lack the ability of "learning to learn" that pigeons, horses, cats, monkeys, and apes all possess. But when rats are trained to discriminate between odors, they do improve. The problem is simply that rats have quite poor vision. Likewise, a careful study of pigeons found that when presented the problem of choosing which of two pictures matches a simultaneously displayed sample, the pigeons have a basic perceptual difficulty making direct visual comparisons.

NATURAL CATEGORIES

The ability to master relationships—identity, same-color, same-shape, sameness in general, difference in general, conditionality—would seem to clearly reflect a level of cognitive processing at work. But these tests tell us almost nothing about what exactly is going inside an animal's mind. Does an animal have some literal mental picture of these concepts? Is it applying some sort of analogical reasoning? Or to take the other extreme, is this really making much out of very little, and perhaps this apparent ability to generalize from experience reflects nothing more than the brain's basic wiring? Perhaps once learning pathways in the brain are formed they can be reused for new, similar problems.

One way to probe more deeply into animals' mastery of relational concepts is by testing their ability to recognize categories of objects. Pigeons have been trained to pick out photographs containing a particular person from a set of photos containing a variety of people; they can distinguish pictures of trees from pictures without trees; pictures with pigeons from pictures without pigeons; pictures with people from those without people. They can reliably tell the difference between aerial photographs that contain man-made structures and those that do not; they can tell the letter A from the numeral 2, even when the characters are presented in novel fonts; they can tell pictures of kingfishers from pictures of other birds; and they can classify pictures into four categories (cats, cars, flowers, chairs). In this last test the pigeons were repeatedly trained with a set of slides until they could categorize these images correctly 80 percent of the time by pecking the correct one of four corresponding keys; then they were shown forty entirely new slides (ten each of cats, cars, flowers, and chairs they had never seen before) to see if they had grasped the general concepts of catness, carness, flowerness, and chairness. The pigeons were able to correctly categorize these new images with a 64 percent success rate.

How do they do it? The obvious answer is that they have formed a mental concept or prototype. The have learned what a cat *is* and are demonstrating that knowledge when they peck the "cat" key in response to a photo of a cat they have never seen before.

The unobvious answer is that they are cheating. As many studies have shown, there are, alas, many ways to cheat on such experiments. The toughest sort of cheating to eliminate is the kind that is made possible by background cues in the photos that the experimenters themselves are unaware of.

Earlier we encountered the disconcerting example of the capuchin monkeys who seemed to have a flair for distinguishing the category of "people" from "nonpeople," but actually were making their choices according to whether the slides had patches of red somewhere in the image. By pure chance, sorting this particular batch of slides according to the red/no-red rule roughly correlated with the people/nonpeople categories. Given the huge number of background features contained in most normal photographs of natural scenes, this may be a problem of huge proportions in interpreting categorization studies that use natural backgrounds. Maybe it's red patches; maybe it's green patches; maybe it's a certain overall brightness; maybe it's circular features of a certain shape; maybe it's trees; maybe it's hills—the list goes on and on. With so many irrelevant cues to work from, *one* of these, just by chance, may well turn out to appear more often in the category-X slides than in the non-category-X slides. And it is in the interests of the monkey to "find" a pattern in the slides that gives him rewards, the most efficient way possible. If there is a simple pattern to background cues that roughly correlates with the "true" categorization, it may well be easier for the monkey to glom on to that through a simple learned association.

The problem for the experimenters is: How do you control for the presence of cues that you are not even aware of yourself? One way is to train a second group of subjects to learn a "quasi-concept": The people and nonpeople photos are divvied up randomly into two packs; the monkeys are then rewarded for pushing a key whenever a picture in the first pack is shown. If the quasi-concept is harder to learn than the people concept, then—unless we're extremely unlucky in the matter of irrelevant background cues—it is pretty certain that the monkeys in the first group really are using the presence or absence of people to make their choices.

This test was applied to a remarkable experiment in which pigeons were trained to distinguish underwater photos containing fish from those without fish. Fish were deliberately chosen because they are not something in the realm of a pigeon's normal experience, nor its recent evolutionary history. The pigeons did quite well at the task, and even did fairly well when challenged with tricky cases such as photos of turtles or SCUBA divers. The quasi-concept test effectively ruled out the idea that they were basing their answers on inadvertent background cues that happened to correlate with the fish/nonfish categories.

But there are other ways to "cheat" that are even harder to sort out. The scientists who taught pigeons to categorize As and 2s claimed in their account of this experiment that pigeons had learned the "concept" of an A. Yet subsequent testing with the entire alphabet showed that the pigeons readily categorized as As other letters that have an apex at the top and two projections at the bottom (such as N), and they categorized as 2s letters with curves (such as S). They had not learned a concept of an A or a 2 as a prototype or mental image at all; they had learned to make a visual discrimination between pointy shapes and curvy shapes.

The two statements carry hugely different implications, and they illustrate well the dangers of loose terminology that Vidal and Vauclair have warned against; it is all too easy to "up the ante" by describing an animal's feat in human terms. Tests in which pigeons learned to discriminate pictures with people from pictures without people don't tell us that the pigeons have formed "a concept of people"; the concept may instead be mammals, or animals, or things with faces, or things with legs, or it may be nothing that can be described in words at all.

HOW PIGEONS, MONKEYS, AND PEOPLE SEE THE WORLD

Even assuming that the animals have formed a conceptual category, there are no guarantees that their categories have anything to do with the ones we see in the same pictures.

One particularly revealing experiment, because it had pigeons, monkeys, and undergraduate students all perform the same task, showed how difficult it is to know for sure how an animal is drawing its distinctions. In the first part of the test, the subjects were shown two pictures at a time; one picture was of a kingfisher, the other of some other bird. Touching the screen with the kingfisher yielded a reward (banana pellets for monkeys, grain for pigeons, and the sounding of a tone for the apparently easily entertained undergraduates). Once this task was mastered, the subjects were then shown novel pictures of king-fishers versus other birds. Pigeons, monkeys, and people all did fairly well in transferring their training to the novel pictures. But particularly interesting was that almost none of the people who were asked the basis for their choices realized that the pictures that triggered the "correct" indication were all of the same species of bird. Most of the humans said they had just followed a strategy of picking the picture of the more col-orful bird. When the set of novel pictures was redone to pair the king-fishers with other colorful birds, performance of humans dropped substantially, from about 94 percent correct to 77 percent correct. The monkeys and pigeons did worse, too, dropping by about 10 percentage points, from 80 percent or so right to 70 percent.

So really the subjects were not so much forming a general category of kingfishers as they were just cueing in on a single and easily distin-guished variable—colorfulness. If an animal perceives all members of category X as identical—either because a simple rule (e.g., "colorful") adequately distinguishes them from the items presented in a contrast-ing category, or because the items within the category simply do not vary much from one to another—then the animal has not performed any task of categorization at all.

That might seem to be philosophical hairsplitting, but it actually goes to a quite fundamental point about these studies. When a group of different items elicit an identical response from an animal because the an-imal perceives them *to be* identical, we haven't learned a thing about the animal's ability to form mental categories and generalizations. The pi-geons that learned to distinguish As from 2s regardless of font were ar-guably not performing any feat of generalization at all. They picked out one (or at most two) salient features of the objects and used that as the basis for all of their discriminations. One could argue that as far as the pigeon was concerned, a Times Roman "A" and a Bodoni "A" didn't just look categorically similar—they looked identical.

With some clever further testing, though, it is possible to determine if in such cases the animals are in truth forming categories or if they are just making simple visual discriminations. In an experiment with baboons, Jacques Vauclair and Joël Fagot first trained the monkeys to distinguish Bs from 3s in a variety of fonts, then challenged them to categorize Bs and 3s in novel fonts. The baboons did well, just as the pigeons had in the earlier experiments. But then they tried a second test to make sure that the baboons actually were able to discriminate between individual items within each category—that is, whether the monkeys could even tell the difference between a B in one font and a B in another. This they did by first training the baboons to perform a matching-to-sample test with random geometric forms. Then the baboons were presented with a new matching-to-sample test: the sample was a B or a 3, and the baboons had to choose between a character that exactly matched the sample (the right choice) and one that was the same character but in a different font. This test showed that the monkeys were indeed able to tell the difference between characters in different fonts—and so must have made a true generalization in order to have placed them into categories.

The pigeons and the baboons, in other words, performed two tasks that looked almost identical but which probably involved completely different cognitive processes. This is a wonderful illustration of why practical tests of animal cognition can never wander very far away from philosophical issues.

Only when we are sure that an animal is actually forming categories can we start to tackle the really interesting question of how their minds do it. One theory is that animals form a single, Platonic prototype of the category, to which they then compare novel images. Morris the Cat looks more like the Ideal Cat than he looks like the Ideal Chair or the Ideal Car; thus the pigeon concludes that Morris is a cat. A variation on this theory is that animals recognize members of a category by storing an entire library of exemplars for each. A pigeon might retain a mental representation of every person and nonperson picture it was shown; when presented with a novel image, it would in effect run through its whole library of images and see which batch—persons or nonpersons—the new image most closely resembles.

A competing theory holds that animals form groups by maintaining stripped-down lists of rules governing "family resemblance." Not every feature on the list would need to be present in every case; a pigeon might categorize an object as a tree if it contains several of these characteristics, or some established weighted average among them, which the pigeon has shaped and refined as it encounters new examples. These characteristics might include greenness, leaves, branching limbs, and an overall shape that rises vertically from the ground.

A number of experiments lend credence to the "family resemblance" theory. In the kingfisher/bird study, subjects were also trained to distinguish bird/other-animal categories and animal/nonanimal categories. The monkeys and pigeons had a surprisingly hard time mastering either of these. With extensive training they did eventually perform fairly well at categorizing novel pictures in the animal/nonanimal test, but never as well as they did on the kingfisher/other-bird test. Performance on the bird/other-animal test, meanwhile, never exceeded chance.

Studies in rhesus macaques showed a similar phenomenon: the monkeys could learn to distinguish pictures of rhesus macaques from Japanese macaques, but could not learn the more general categories of all macaques versus other monkeys.

The prototype and exemplar models of category formation would be hard pressed to explain these results. Subjects shown a lot of bird pictures ought to have built up a substantial library to make comparisons with. (Some animals in that experiment were trained repeatedly with 250 bird/other-animal pairs, yet still were unable to make accurate discriminations of novel photo pairs.) A novel bird image certainly ought to more closely resemble a library of bird pictures than a library of other-animal pictures. Moreover, because the images in the bird library more closely resemble one another than do the images in the animal library, it ought to be easier according to these theories to pick out a bird (a narrowly defined group) from a sea of animal images than to pick out an animal (a broadly defined group) from a sea of images of all sorts of things. Yet all the animals who were tested found it harder to do so.

The family-resemblance model does a better job of explaining these results. The rules for distinguishing an animal from a nonanimal ought to be fairly simple; it's an all-or-nothing choice. But the rules for telling a bird from a nonbird animal would be much more involved, given how many features birds from the start share with other animals—legs, eyes, mouths, and so on. Picking a specific species from others again becomes a relatively straightforward proposition according to the family-resemblance model; a species has unique distinguishing characteristics that hardly demand generalization at all. The fish-recognition researchers, R. J. Herrnstein and Peter de Villiers, note that it is easy to recognize a particular Italian man in a series of pictures of people; it is easy to recognize men in general; but it's hard to recognize Italian men in general. The middle levels of categorization are the hardest. Indeed, recognizing an individual may not even involve "categorization" at all; if the range of items in the alternative category choice is limited, it's possible to get more or less the right answers just by cheating. When the pigeons had only to choose between As and 2s, they could get by with looking at nothing more than the apex of the A and curviness of the 2; when people had to choose between kingfishers and dull birds, they could do so reasonably well just by comparing the colorfulness of the birds.

ENCODING THE RULES

How might these rules, or some weighted average of such rules, be represented in an animal's mind? People sometimes can articulate after the fact how they have formed categories, but the actual process often appears to take place quite unconsciously. And sometimes people really cannot explain what they've done, just as children often seem to be able to master the rules of a new game by watching others play but cannot put the rules into words. It seems unlikely that a list of rules for categorizing objects is somehow recorded and executed sequentially in the brain. One possible explanation invokes an analogy to artificial neural networks: Imagine you have a array of photocells, each of which can detect light, wired up so they converge on a network of higher level "nodes"—just as the simple light-sensitive cells in the cat's retina feed into layers of complex and hypercomplex cells. You have a panel of knobs that can adjust the relative strength of the connections between layers. You "show" the photocell array a picture of a cat and twiddle the knobs so the signal at one of the top-level nodes is maximized. You then show it a picture of a fish and twiddle the knobs a bit more to maximize the output at a different top-level node. If you keep doing this with more and more pictures of cats and fish, you eventually achieve a network that can do a fairly good job of discriminating novel images. Show it a cat it has never seen before, and you get a signal at one node, show it a fish and you get a signal at the other. If you showed it a lobster, you might get a weak signal at both; show it a dog and you would probably get a stronger signal at the cat node than the fish node, but not as strong as a cat image would evoke.

Is the network applying "rules"? Well, yes and no. They probably aren't rules that can be articulated with any linguistic or logical precision. The rules exist in a form that is distributed throughout the connections of the network. The act of twiddling the knobs automatically adjusts the relative weights that are given various features of the images; the choice of what features the network pays attention to is an operational one and cannot be broken down to anything as discrete as "legs" or "furriness."

How well do artificial neural networks duplicate what real animals do? Like neural networks, the rules that animals seem to use defy precise articulation. In the animal/nonanimal experiment, the researchers tried to analyze whether particular animal photos had specific features that made them harder or easier to identify, and mainly came up dry. They found almost no differences between the pictures that the pigeons and monkeys had an easy time identifying as animals, and those they had a hard time identifying. The easy and hard pictures were indistinguishable in terms of how much of the animal was shown, how big it was, how colorful it was, how closely it matched an "ideal" concept of an animal. In other words, both animals and artificial neural networks appear to categorize pictures based on properties of the image as a whole. (The one significant—and negative—correlation they found was that the pigeons and monkeys tended to do worse on pictures where the animal's eyes were prominent. Because most animals tend to react to another animal's eyes as a threat, the subjects may have tended to avoid these pictures and so tended to misclassify them.)

There are, however, significant differences between animals and artificial neural networks in actual performance. An artificial network did relatively well in learning to "identify" the comic-strip character Charlie Brown, but relatively poorly at real people in natural settings; pigeons were just the opposite. Pigeons are not terribly good at abstract categories such as colored lights; they do better with natural categories such as fish; and they do best with natural categories that have evolutionary significance to them, such as trees, leaves, and people. That in itself does not rule out the idea that animals basically represent categories in their minds through the strength of interneural connections starting at the retina. But it does imply that, rather than a sort of general-purpose network that can be adjusted in any direction by twiddling the knobs, the animal mind has had some perceptual biases that have been built into it over the course of evolution. Recall that some cells in the monkey's visual cortex seem to respond preferentially to a shape of a monkey's hand; it would hardly be surprising that at least some of the wiring in the visual processing hardware of a brain is biased toward other important shapes that appear in the real world of its owner. Cats, which after all are carnivores that need to detect moving prey, have been found to be able to distinguish between "biological" and "nonbiological" motion: They were shown fourteen dots of computer-generated light that either formed the outline of a walking or running cat, or were scrambled versions of that same representation. The cats consistently could tell the two sorts of motion apart. They could not distinguish between the two when the biological motion display was turned upside down. Other species likewise show a special bent toward classifying objects that are of immediate biological importance to themselves. Blue jays, for example, are able to categorize slides of cryptic moths that would probably stump most people.

Even though fish are not a part of the pigeon's environment, and have not been for millions of years of evolution, they are biological and share some general traits with biological objects that are of more immediate relevance to a pigeon's life. A neural network optimized to categorize sources of food or sources of danger in the real world may, by its very wiring patterns, do fairly well on other natural objects, too—and fairly badly on line drawings.

MAKING LISTS

If you were supplied with a computer, a more or less normal person, and an extraordinarily accurate timer, you would be able to conclude in very short order that people and computers do not go about multiplying large numbers the same way. Whether you tell a computer to multiply 3×3 or 300×3000 or 0.3×0.003 makes no difference; the amount of time it takes to spit out the answers will be identical. People, though, usually take much longer to do the second or third problems. Although few people have trouble with 3×3, the only actual multiplication task in all three of the problems, most people find they need to carry down all the zeroes to make sure they get the decimal point in the right place. Computers on the other hand do that job automatically because of the way they store numbers in the first place. Every number that goes into a computer gets converted to a representation known as scientific notation, which consists of a decimal number between 1 and 10 multiplied by a power of ten. So 300 is represented as 3×10^2, 3 as 3×10^0; 0.3 as 3×10^{-1}. Whenever a computer multiplies two numbers, it just adds together their powers of ten to figure out where the decimal point goes. So in all three problems the complexity of the calculation is exactly the same. In every case the computer multiplies 3×3 and then just adds the exponents of 10 together. (For example, 300 is 3×10^2, 3000 is 3×10^3, so in the case of 300×3000 the power of ten in the answer is $2 + 3 = 5$; thus the complete answer is 9×10^5.)

On the surface computers and people appear to have performed the same calculation. But even if we knew nothing about how computers or people actually work, the timing of their performances alone reveal that there must be some underlying, "cognitive" differences.

We could probe further by trying variations on the problems. If it turned out that adding a zero to the larger of the two numbers always increases by a fixed increment the time it takes a person to come up with the answer, we probably could deduce without too much trouble what "algorithm" the person is using. If we notice that the computer makes small rounding-off errors when we feed it numbers with more than seven digits (it treats 12,345,678 as if it were 12,345,680, say)— but that it makes precisely the same errors with numbers such as 12,345,678,000 or 12,345.678 or 0.000000000012345678—we could deduce that the computer is probably representing numbers internally in scientific notation.

So seeing not only *if* an animal can do a task that demands cognitive processes, but also how long it takes to do variations on that task, and what sorts of errors it makes when it does so, is a powerful way to probe what is going on in an animal's mind.

One particularly revealing series of experiments has studied the ability of pigeons and monkeys to memorize sequential lists of items (these studies, of course, are the basis for the ignition lock that the cognitivist friends of B. F. devised). The items might be colors, or geometric shapes, or color photographs of natural objects like tomatoes, weasels, rocks, people, and mountains. Pigeons have been able to learn five-item lists; rhesus monkeys have mastered six items. In the typical experimental setup, the items all appear on a touch-sensitive screen in a random order, and the animal has to press or peck the items in the correct sequence that it has been taught. The animal gets no intermediate hints that it is on the right track; only after the complete series is entered is a reward issued. The animal is first taught the correct sequence in phases. First it sees a screen with nothing but the first item (which we can call item A). After it has learned that pushing A earns a reward, it advances to a screen where A and B appear and the animal is rewarded for pushing A and B in the correct sequence (and not BA, AA, or BB); then C is introduced and the animal is rewarded for pressing A followed by B followed by C; and so on.

Monkeys learn faster than pigeons. But much more interesting is the fact that the two species apparently memorized the lists using fundamentally different internal representations. Herbert Terrace and his colleagues at Columbia University were able to tease out this fact through some clever variations on the experimental theme. Once the animals were able to correctly play back the list they had been taught, they were presented with a modified task. Instead of seeing a screen containing all of the items in the list, they were shown a subset containing just two items. Their task was to press them in the same order they appeared in the complete sequence; for example, when shown a screen containing just item A and item D, they had to press A first, then D. And here is where strange things started to happen. An animal that possesses a complete "picture" in its mind of the list ought to have no trouble with this problem. But the pigeons showed some very odd errors. Whenever the subset contained A or D (in the case of a four-item list), they did fine; AB, AC, AD, BD, CD all produced high scores. But when challenged with the subset BC, the pigeons never scored better than chance.

This was odd on two counts. When given the task of producing the entire four-item sequence, the pigeons had no trouble pressing C after they had pressed B; yet they could not do it when presented with the subset BC alone. This clearly ruled out the possibility of the pigeons having formed a complete, linear mental representation of the full sequence of items. It also ruled out what might seem the simplest procedure for learning a list, a process called "chaining," in which pairwise sequences of adjacent items are memorized—A is followed by B, B by C, C by D. Yet at the very same time, the pigeons were able to do something far beyond what chaining would permit: they could place nonadjacent items (AD, AC, and BD) in the correct order.

Monkeys, by contrast, were able to place all ten subsets of a five-item list in the correct order. The real stroke of insight into what was going on came with an analysis of how long it took the monkeys and pigeons to respond to the various subset tasks. With monkeys, the time it took to push the first item of the pair was almost precisely proportional to how far down the list that first item fell. In subsets containing A as the first item, the monkeys picked out A in about 1¼ seconds; when the first item was B it took about 2½ seconds; C, 3¾ seconds; D, 5 seconds. Likewise the time to choose the second item was proportional to the number of items skipped between the first and second items of the subset; that is, choosing the second item of the pair AB took less time than AC, and AC less time than AD.

These timing analyses strongly suggest that the monkeys were in effect running through a stored list, in order. To find the first item, they began at A, and ran through each item in sequence until they found one that matched an item on the screen.

The pigeons by contrast showed no such patterns in timing. They took almost the same amount of time in every case. So they were not running through a mental list, but must have been applying some other rules. Terrace believes the explanation is that the pigeons learned the original list through a series of linkage rules that are in a sense actually simpler than those involved in chaining—and which reflect the phases that the pigeons went through to learn the list in the first place. In effect the rules are first peck A; after you've pecked A peck B; after you've pecked AB peck C; after you've pecked ABC, peck D. The fact that A appears at the start of the list and is repeated over and over in the course of learning the entire sequence has the effect of making this starting point a particularly salient feature. The last item of the list, D, is also distinctive. Terrace believes the pigeons respond to the subset tasks by applying three rules: If A is present, pick that first; if D is present pick that last; make all other choices by default. So in the case of BD, the pigeon is making but a single decision: pick D last, which forces by default that it pick B first. But these rules are of no help in the "problem" subset of BC, thus the pigeons' chance performance on that one pair.

One lesson from these studies ought to be awfully familiar by now: just because an animal has learned to do perform Task X does not necessarily mean it has a complete mental representation of X stored away in its brain. It would be easy to look at the basic ability of a pigeon to learn and replay five-item lists and assume that it had stored a representation of the entire list in its brain. Only after the further examination of reaction times and subset ordering does it become clear that the pigeons were doing something short of that.

The other lesson, though, is quite exciting—the confirmation that monkeys really have an ability to store items in sequence and mentally run through and draw conclusions from such a sequential list. Other studies have shown that monkeys are able to build up complete and correctly ordered lists from pairwise "chaining," and then run through the complete lists to make correct judgment about the relative order of nonadjacent items. For example, monkeys were trained with rewards for choosing E over D; D over C; C over B; and B over A. When presented with a novel choice like D versus B, they would consistently choose D—even though in their training the choices B and D had both been rewarded with equal frequency. Here is a striking case where a behaviorist model would predict a totally different result from the one obtained. If the monkeys were responding simply to their history of reinforcement, they would show an equal chance of picking either member of novel pairs such as B versus D, or E versus C, or E versus B. Instead, the monkeys spontaneously applied a "transitive" rule to a list they had constructed. Even more interesting is that an analysis of reaction times showed that the monkeys were again running in sequence through the list they had created—and that they consistently started from one end (though which end was used as the starting point differed from individual monkey to individual monkey).

Further confirmation that monkeys learn the actual ordinal position of items in a list comes from a particularly intriguing experiment that Terrace's group performed. The monkeys were first trained to pick four simultaneously displayed photographs in a specific order—for example: elk, rock, leaves, person. It typically took the monkeys 400 or more trials to master each list to the point that they could reproduce it correctly 75 percent of the time. Once the monkeys had mastered four such lists, they were given several different types of new lists to learn. Some of the lists drew one item from each of the four previously learned lists, but kept them in the same ordinal position as they had appeared originally. So if elk was the first item in the new list, it was also the first item in one of the previously learned lists; the second item in the new list would be the second item from a different list; and so on. In these new lists that maintained the items each in their previously learned ordinal positions, the monkeys took about 100 trials to learn. But when the list placed the items in a sequence that took each item out of its previously learned ordinal position, the monkeys took essentially as many trials to learn as totally novel lists. The monkeys had clearly maintained a representation of the ordinal position of each item which they were able to transfer efficiently to new problems.

COUNTING

In a number of animal species that breed repeatedly in one season, the females make a ruthless—though mathematically impressive—calculation. If the brood size is too large, the female may cannibalize some of her offspring. If the brood is too small she may abandon it altogether and rebreed immediately, a sound evolutionary strategy if in so doing she increases the average number of offspring successfully raised per unit time. Birds similarly seem to favor a predetermined number of eggs in a clutch. If you swipe eggs from the nest of a laying bird, it usually lays more. If extra eggs are sneaked into the nest, the bird may stop laying. In all of these cases the observed behaviors must be triggered by some numerical judgment the animals are making.

Counting would seem the perfect subject for a study of animal cognition. The use of numbers is, after all, an example of nonlinguistic thinking par excellence. There is no doubt that many animals can make numerical judgments. But beyond that general statement the situation gets extremely murky, in no small part because of the anthropocentric bias we bring to the whole area of mathematical ability. One thing that the many studies of numerical ability in animals have pointed out is how many ways there are to use numbers that are very different from what humans have learned to do with the consciously employed tools of formal mathematics.

Even so simple a question as whether birds are able to count their eggs immediately leads us into philosophical bogs filled with thorns and gloom and semantic debates over the difference between "numerosity" and "numerousness" and "numerons" and "numerlogs." Even worse is the discovery that there is probably no path that skirts the bog. The best we can do is take a metaphorical helicopter ride over this territory and try to pick out the most important landmarks.

A number of other species have done well in laboratory tests in which they must make correct relative or absolute numerical judgments. Raccoons, for example, were presented with five clear Plexiglas cubes containing from one to five grapes and were taught that only if they picked the cube with three grapes would they be allowed to eat them. The training period was by no means brief, but once the raccoons finally were able to reliably pick the three-grape cube, they readily transferred that ability to picking out cubes containing three raisins or even three toy bells. Rats have been trained to choose the fourth tunnel along a wall, even when the spacing between the tunnels was changed repeatedly. Pigeons can be trained to peck at whichever of two screens has more dots (or more people); they can even do this when the objects are not the same on the two screens, for example pictures of flowers versus pictures of birds, or Xs on one screen and Os on the other, or big dots on one screen and little dots on the other.

Most people would say that the animals in each of these cases are counting. But a closer look at their performance—and at the kind of mistakes they make—point to simpler cognitive processes that may be at work. An important point about human number use is that people not only can count in order, but also understand general concepts about the relationships between numbers and what they stand for that go beyond counting. A seven-year-old child can tell you whether 173 is bigger than 142 without having to count from 1. Older children explicitly grasp the notion that counting implies the possibility of extending the process indefinitely, but even young children do not need to be taught each number separately; a child who can count to 24 doesn't need to be taught that there are numbers called 25 and 26 and shown pictures containing 25 or 26 items to get this idea across. If an animal is truly counting, there ought to be no limit to how high it can count.

So what are animals doing? As early as 1871, psychological researchers noted that humans can make very fast and accurate assessments of how many objects are present in small arrays. If you are shown a dish with three raisins and a dish with three cookies, you don't need to count them to see at a glance that each contains the same number, and that that number is three. This process of recognition is probably a perceptual one: An array of three objects presents a characteristic visual appearance—a "threeness"—in much the same way that cows or birds have characteristic features that we can perceive and correctly categorize at a glance. This perceptual process, sometimes called "subitizing," is quite effective up to numbers of around six for most people. Even if you could not count and had no idea what a number was, you could reliably distinguish between groups of three and groups of four things by this method—though you might just as well call these two categories of things Fred and Adele or Rocky and Bullwinkle as call them three and four. People who *can* count are able to discover that "threeness" perceived in this fashion is in fact equivalent to the absolute number 3 you get by counting. There is some evidence (though this claim is not free from controversy) that children learn the names of the first few numbers—without a developed concept of numbers or of counting—by attaching labels to arrays perceived by subitizing; only later do they learn to connect these same names to the process of counting.

An obvious limit to how far the ability to recognize quantities in this fashion can be stretched is that, as the number of items increases, so does the number of different ways they can be arranged. Three items can be placed in a straight line or a triangle, and that is about it. Four items can be arranged in line, a square, an L shape, a T shape, a Y shape, and so on. The possibilities increase dramatically with each successive number. There is nothing about the visual appearance of twelve items that can reliably distinguish them from thirteen items.

The raccoons taught to pick the box with three things could well be subitizing; so could the several apes who have been trained to correctly identify the number of objects in a set by picking from a series of buttons or cards labeled with Arabic numerals. Of course there is something just a bit fishy about using Arabic numerals in such experiments. If all you want to do is see whether an animal can recognize and distinguish between arrays containing one through six items, you could train the animal to use any arbitrary symbols (colors, shapes, letters); you could, for that matter, teach them to use the Arabic numeral 6 to stand for three, 2 to stand for five, and so on. Seeing a chimpanzee use the same numbers we do inevitably has the effect of suggesting that more is going on than may be the case.

There are other ways people, and possibly animals, use to represent numbers that don't involve counting, and these methods may work for much larger numbers. One is to take advantage of rhythmic patterns that often come along with numbers when they appear in the natural world. As Hank Davis and Rachelle Pérusse have pointed out, most people can repeat the nonsense refrain of the Christmas carol "Deck the Halls" (fa la la la la la la la la) without any trouble at all; almost no one can tell you, without stopping to count, that there are eight la's there. And note that because they are repeated nonsense syllables, there are no mnemonic clues that help, the way the syntactically related words of a poem help us. Rhythmic representations are very hard to rule out in explaining the results of many animal experiments that seem to show counting ability.

There may be other sorts of mental representations that accomplish the equivalent of counting that have nothing whatever to do with numerical processes. We have already seen that many animals can learn list sequences of the kind A, then B, then C, then D. For example, a pigeon can be taught to peck at four photographs in a specific order. When a rat is trained to pick the fourth tunnel, is it counting the way we would—or is it executing a learned list sequence that instead of taking the form ABCD is taking the form AAAA? If we recall the discovery of the "magic number seven," it certainly is more than possible that in learning to do four or six identical things in a row, an animal is simply using its working memory to create a list and not actually "counting" at all, but merely making comparisons exactly as it would do in running through a list of different items. When a rat can pick the twenty-third tunnel from the twenty-second or twenty-fourth, then we may be getting somewhere.

Finally, there is a whole class of perceptual abilities that fall under the rubric of "relative numerousness judgments." Many animals show a natural preference for greater quantities of food items over smaller quantities. There are many cues nature offers to aid in such visual discriminations—cues that obviate any need for a sense of absolute quantity. Most obviously, animals can cue in on the total area covered by the objects—you don't have to count the number of birdseeds in a pile to determine that the pile that simply looks bigger has more of them.

If it so happens that the design of an experiment does not take pains to eliminate such cues, it is impossible to tell if the animal is counting or making a relative numerousness judgment. If the objects are of uniform size, then a slide showing more objects will appear to simply have a bigger "blob" of stuff. If the slides are shot with a uniform bright background, then the one with more objects will have more of the background covered and thus a lower overall brightness. If the size and spacing of the objects is varied so that the total area covered in both slides is the same, regardless of how many objects each has, then another problem arises—the spacing between the objects itself becomes a cue. In the slide with fewer objects, each object is farther apart from its neighbors. Such cues are maddeningly hard to control for. In a way, though, controlling for them misses the whole point. Whenever pigeons can distinguish sixteen dots from nine dots, they probably *are* using physical cues that permit a relative numerousness judgment. One thing for certain is that they are not counting or subitizing. While pigeons can apparently tell twenty-five from sixteen and twenty-five from thirty-six, they cannot tell two or four from three or five.

CAN AN APE ADD?

Many claims for numerical knowledge in animals have been based on their ability to master ordinal relationships. In one of the more impressive studies of this genre, Rhesus monkeys were taught to associate Arabic numerals displayed on a computer screen with varying food rewards. The screen would display a pair of Arabic numerals from 1 to 9; when the monkey moved a cursor onto one of the numerals, he would receive that number of monkey pellets. In time the monkeys learned to reliably choose the larger number—even when they were presented with a pair of numerals they had never before had to choose between. So, in other words, they were not just memorizing the right answers to all forty-five possible different two-number combinations (e.g., choose 7 when the choices are 7 and 6; choose 7 when the choices are 7 and 5; and so on and so forth): They were applying a transitive rule (for example, if 7 is better than 6, and 6 is better than 5, then 7 is better than 5). In another test, the trained monkeys were shown a screen with five different numerals; the monkeys would reliably "count down," choosing the largest number first and so working their way down to the smallest.

But there are some interesting methodological (and logical) questions about what is going on here. We have already seen that monkeys and other species are able to form sequential lists. The fact that in this experiment Arabic numerals were used as tags adds nothing to that accomplishment. Being able to store nine different symbols and place them in a transitive sequence is certainly an impressive feat. But the real question about numerical competence in this experiment comes down to whether a monkey has to actually count to tell the difference between nine monkey pellets and eight monkey pellets. What the monkey may be reacting to is purely the relative "hedonic" value of each symbol (more food feels better), rather than a concept of number at all. That is a question this experiment does not directly answer. Again we are back to the issue of counting versus relative numerousness judgments.

Another experiment that has stirred much interest and speculation involved the chimpanzee named Sheba, and it seemed to show an ability to use the abstract concept of numbers in a representational manner. Sheba was first taught to associate a tray containing one, two, or three gumdrops or M&Ms with placards containing the corresponding number of metal discs. Sheba would be shown the tray of candies; if she picked the correct placard she would get to eat the candies. Next the placards were replaced with placards bearing the Arabic numerals 1, 2, or 3. Finally, the candies were replaced with inedible objects like flashlight batteries or spools; if Sheba chose the numeral that matched the number of objects, she would be rewarded with the corresponding number of candies. Along the way the numerals 0 and 4 were added to her repertoire.

So at this point Sheba was able to indicate the number of objects in a tray by pointing to the corresponding Arabic numeral. So far everything she had done was consistent with subitizing. As with other such experiments, the use of Arabic numerals carries a slightly misleading suggestion, for there was absolutely nothing to suggest that her ability to recognize and distinguish one, two, three, or four objects was in any way related to a true concept of absolute numbers.

The training phase complete, Sheba was then given a new challenge. Instead of being presented with a single tray full of objects, the objects (oranges in the first tests) were spread out at two different sites in the room, neither of which was visible from the other or from a platform where the numeral placards were kept. Sheba was led to the orange caches, then to the platform where she was supposed to pick the right placard that represented the total number of oranges. Overall she got the right answer about three-fourths of the time, far above chance. Finally, the oranges were replaced by placards bearing numerals that Sheba had to add, and again Sheba got the right answer about three-fourths of the time.

Adding up absolute quantities, and even more so using numerals to stand in for the absolute quantities, cannot be readily explained by subitizing. Sheba seemed to have an ability to call on some absolute quantitative representation of an Arabic numeral, combine that with an absolute quantitative representation of another Arabic numeral, and convert the resulting quantity, now contained only in her head and not in any physical array before her, to another Arabic numeral.

But the relatively small number of combinations that Sheba had to tackle to produce these impressive results really ought to raise suspicions. In the first of the adding-the-oranges problems, the sums were restricted to one, two, or three. There are only five different possible combinations of oranges that can add up to these figures ($0 + 1, 0 + 2, 0 + 3, 1 + 1, 1 + 2$). Sheba was given eighty-two trials on these problems, which means that each sum was repeated about sixteen times on average.

Now consider this parallel experiment conducted by that useful if never particularly likable personage, the devil's advocate. Our hypothetical subject, a hypothetical chimpanzee whom we will call Solomon, is brought directly into the room with the concealed placards. Instead of bearing the Arabic numerals 1 through 4, however, the placards are red, orange, yellow, and blue. Solomon, like Sheba, is led round and shown the two concealed placards, then brought back to a platform where he has to choose an orange, yellow, or blue placard. We give him eighty-two attempts to learn these five rules:

Color of concealed placards		*Color of placard to choose*
red and orange	⟶	orange
red and yellow	⟶	yellow
red and blue	⟶	blue
orange and orange	⟶	yellow
orange and yellow	⟶	blue

Now, would we call this counting? In fact this problem is formally identical to Sheba's accomplishment; we have simply substituted red for 0, orange for I, yellow for 2, and blue for 3. Chimpanzees are well known to be able to learn new discriminations in a single reinforced trial. Let us assume that Solomon, like Sheba, has already had extensive prior training in discriminations, same-difference concepts, color discrimination, the use of colors as attributes, and one-to-one correspondence. Note that three of the rules are particularly simple and share a common conditional principle of the sort that chimpanzees have no trouble learning: if one of the items is a red, then pick the placard that corresponds to the other item. Note also that there are only five problems and three possible answers to each. Suppose Solomon adopts this straightforward strategy: he tries answers at random to each problem until he hits on the right answer and gets a small reward—thereupon learning the rule for that problem. A straightforward probabilistic calculation reveals that there is a 95 percent chance that he will have solved all five problems after making a total of twenty or fewer wrong answers—which works out to a total score of at least three-fourths correct on eighty-two trials.

Like a young child who memorizes a story and then pretends to read it, Solomon was simply learning by rote the correct answers he discovered by trial and error. He was not doing math any more than the child is really reading. By following a purely random-guessing strategy he was almost certain to produce as impressive a score as Sheba achieved. Perhaps there is a lesson here for Sheba's performance, too. And perhaps if Sheba's placards had not been labeled so compellingly with Arabic numerals, we would not have been so quick to equate what she did with something mathematical.

The larger point is that the very limited numerical abilities that even extensively trained higher primates show has to raise doubts. Sarah Boysen, Sheba's trainer, herself acknowledges that it required "a heroic effort" to obtain what numerical performances she did from Sheba. A true concept of number ought to be infinitely extendible and infinitely transferable. Yet there have been no convincing studies that an animal finds it easy or automatic to transfer numerical performance from even one perceptual realm to another. It is no easier to train an animal to push a key twice in response to two beeps and four times in response to four beeps than the other way around.

Even more to the point is Hank Davis's warning to study what animals are naturally good at: "Absolute numerosity is a distinctly human invention. No nonhuman animal needs this form of numerical competence to lead a successful, totally normal life. . . . Once again human investigators have taken some characteristic human ability and gone looking for rudiments of it in other species. Certainly, there is arrogance or anthropocentrism to this activity."

5

MAPS, TOOLS, AND NESTS

MY HORSE WOULD MAKE AN EXCELLENT MEMBER OF the local historic preservation society, for he unfailingly reacts with indignation to the least alteration in the built environment.

If someone along our road puts a new piece of lawn furniture out in the yard, or replaces their mailbox with a bigger one, or cuts down a tree, or puts up a new fence, my horse will, upon catching sight of this change, stop short, prick up his ears, stare hard, snort, and sometimes skitter to the opposite side of the road. The roughest day to take him for a ride down the road is Wednesday—trash day—when trash cans appear at the end of various driveways, and he careens from one side of the road to the other.

From a horse owner's point of view, this sort of domestic animal behavior is merely annoying. And to most people, indeed, it probably does not seem to amount to any particular feat of animal brilliance.

But think what is involved here: The horse has had to acquire a considerable library of mental representations of its world; it must know what point in space each of these "snapshots" corresponds to; and it must be able to call these snapshots up and compare them to the visual image it is currently receiving and spot any discrepancies. This is something animals do day in and day out with barely a nod of recognition from us, yet it surely represents a highly developed cognitive ability.

Even more so the ability to perform complex feats of calculation on such stored spatial memories—something animals also do all the time. A horse not only can become familiar with the scenery along miles of roads and trails, it can also learn which turns to take along an accustomed path with dozens of intersections.

Clark's nutcrackers perform even more prodigious feats of spatial memory. These birds gather piñon pine seeds during late summer and cache them in crevices or underground to feed upon in winter. They may collect tens of thousands of seeds and store them in literally thousands of separate locations. Field observations and experiments appear to have ruled out the possibility that they bury the seeds, forget where, and find the caches later when they need them simply by searching from scratch. (Squirrels do appear to find buried nuts this way; with so many squirrels burying so many nuts, it's not hard to hunt down a cache by smell or by searching for signs of disturbed ground.) In a series of experiments, captive nutcrackers were placed in a ten-by-ten-foot room, supplied with food from a centrally located feeder, and allowed to bury the seeds in sand spread over the floor. When brought back to the room a month later, the birds were able to find their hidden caches at a level significantly greater than chance. When the researchers hid seeds themselves, the nutcrackers did not do nearly so well at finding them. Nor did they do as well when the room was stripped of distinctive "landmarks" such as rocks and logs of various sizes. The birds generally buried their seeds near such landmarks; when the researchers moved the landmarks around during the time the birds were out of the room, the birds had a hard time finding their caches, and would generally be seen to search around one particular part of a log or rock—again strongly suggesting that they remembered specific spatial relationships.

A similar experiment with another caching species, the marsh tit, offers further proof that the birds are not merely rediscovering forgotten caches by smell or other cues, but are actually remembering precise locations in space. The researchers in this experiment took advantage of the convenient fact that the left and right halves of a bird's brain share virtually no visual information with one another. What comes in the left eye is processed and stored by the right brain, and vice versa. In the marsh tit experiments, when the birds had one eye covered throughout, they did just as well as they had done using both eyes. But if the birds were permitted to store the seeds with the left eye covered and then set to retrieving the seeds with the right eye covered, the tits performed only about as well as if they were searching randomly. In other words, in recovering their caches, they clearly relied on visual information processed and stored in the brain while they were choosing their cache sites. Switching eyes half-way through made it impossible for the tits to compare what they were seeing with what they had seen.

FALSE SCENTS

As usual, we have to be careful here. Some of the most impressive feats of animal map-making and navigation turn out to be the most brainless. For many years scientists were baffled by the annual migration of salmon, which find their way upstream to spawn and die in the very place they themselves had hatched. Going downstream, a fish has no navigational choices to make; it is simply swept along with the flow as tributaries feed together. But heading upstream the fish must make a correct turn at every branch. How do salmon do it? Do they remember each turn as they pass by on their way to the sea—even though the turn was not one they had to make?

It finally turned out that there was a far simpler explanation that did not involve maps or remembered branches or spatial memories at all: The fish follow minute but geographically distinctive odor trails flowing down from their place of origin. At each fork, they choose the one tributary that smells like home. Instead of having to construct and memorize a map with dozens of branching-off points, they have learned but a single smell and let the geometric constraints of nature do the rest: water flows downstream and carries smells with it.

Likewise, some other feats of animal navigation that seem particularly impressive (and even mysterious) to us rely not on extraordinary brainpower but on extraordinarily powerful or unusual senses. Some migrating birds have an iron-containing organ in the brain that acts as a built-in magnetic compass. They can fly reliably in a preset compass direction, an ability that is simply beyond our sentient experience. But many such birds have no mental map or spatial memory to go with this compass. If shifted laterally off course, they do not adjust their direction to keep heading toward their habitual wintering spot, but instead continue to fly on the same preset magnetic-compass bearing—and thus wind up displaced laterally from their destination.

Even when animals are unquestionably storing and drawing upon spatial memories, it becomes a tantalizing problem to figure out just what they are doing. The value of getting to the bottom of these cognitive abilities is obvious: Unlike some of the abstract skills that are probed in traditional animal psychology experiments, navigation and spatial memory are abilities that animals unquestionably use in their real worlds. These are evolutionary adaptations to deal with practical problems of survival. They can tell us much about how cognitive processes evolved in animals and what forces have helped to shape the brain and its functions. They are also—and not coincidentally—more complex processes and much harder to pin down.

Consider this seemingly elementary problem: You walk 200 yards along a path out into the middle of a field and place a six-pack of beer on the ground where it is hidden by the grass. You then retrace your steps to your starting point, turn right, and walk 200 yards perpendicular to the path, and stop.

Which way do you go now to find the beer? If you've taken high-school trigonometry, you might calculate your correct course by making a literal map. You know the angle of the path and its distance from the starting point to the beer drop-off point; you know the angle and distance of your subsequent course from the starting point to your current location; you plot the two points, and thereby ascertain the spatial relationship between the beer and your current location.

If you haven't taken high-school trigonometry, you might play it safe and simply retrace your steps along the two legs of the triangle you originally followed. But even if you are a trigonometric dunce, odds are you can strike out on pretty much the correct short-cut course that connects across the two legs.

How do you know how to do that? You might have noticed that

the beer drop-off was near a prominent landmark such as a tall tree, and simply steer toward the tree. Although you're viewing the tree from a different and unfamiliar angle, it still looks pretty much the same; you can tell it is the same tree and off you go. Or you might just have a sense from the distance and direction you've walked that the spot is behind you and to the right—about equal amounts behind you and to the right since you walked about the same distance on each leg of your journey back from the drop-off point. Or you might have a general sense of the shape and size of the field, and have noted that the beer is about a third of the way across the length of the field from its right edge and about half way across its width.

None of these methods is really "mapping" in the strict sense. To make a true map, one needs to survey an area and plot all of the objects according to their geometrical relationships to one another. Making a true "mental map" clearly offers a great advantage over any of these alternative schemes. A complete map allows you to figure out a short cut between any two points, even if you have never actually taken that short cut before. A map allows you to know the relationships between important places without having to travel every possible route connecting them. It allows you to head in the right direction even when your goal is obscured by terrain. It allows you to locate yourself with reference to landmarks you can see.

One way the brain might start to build up a complete map of a landscape is by scanning for distinctive terrain features from a convenient vantage point. The brain calculates the direction that a landmark lies in by monitoring the angle the eyes and head move to observe it; when combined with an estimate of the object's distance, the brain can place it at an approximate map coordinate. To place other landmarks not visible from that initial vantage point, the brain would have to be able to keep track of how far and in what direction its owner travels to reach a second vantage point, then combine that displacement with the eyeball's reports on the angle and distance of landmarks espied from this new location.

The brain is known to be able to keep track of eye directions as it assembles a local picture of a scene, and animals have been shown to be able to estimate by "dead reckoning" how far and in what direction they have traveled. But it is not clear whether any animals that lack symbolic language—including human toddlers—actually navigate by surveying and mapping in the fashion just described. And as our hypothetical beer hunt has already suggested, there are several alternative strategies that animals might use to find their way around.

TRAVELING BY LANDMARKS

Perhaps the simplest strategy is the use of landmarks. Many experiments have clearly shown that animals can remember and recognize objects in their environment. My horse's reaction to things that are out of place along a familiar route is indirect confirmation of his ability to remember miles of familiar scenery. More-rigorous experiments confirm that animals in familiar terrain can be extremely adept at spotting and investigating novel items. A troop of fourteen baboons let into their eighth-of-an-acre outdoor enclosure after a new object had been planted in their absence each day generally took less than three minutes to find it. The baboons clearly reacted differently to the new objects (which included both artificial and natural things, such as drinking cups and coconut shells); they were, for example, much more likely to touch and handle today's new item, while even yesterday's new item quickly joined the roster of same old stuff that they paid only cursory attention to. Similar tests in other species show that moving familiar objects into a new geometric arrangement will prompt animals to reexplore them; otherwise the animals ignore them.

In practice, steering by a series of landmarks is a simple but quite effective procedure. This form of navigation is sometimes called *pilotage;* an animal basically learns from experience that turning right at the rock and then left at the clump of three trees leads home. This way an animal can assemble a catalogue of paths leading to various goals—but still no overall map of the region. This is like the strip of driving directions in a "TRIP-TIK" booklet. Your route is laid out in one dimension, a linear list of where to turn, but it contains no two-dimensional information on how your route relates geometrically to other routes that you may travel on or other sites you may visit in the same region.

An ability to recognize a landmark from several different directions may make it possible to find one's way to a familiar goal or waypoint even when approaching from an unfamiliar direction. The sort of spatial memory that nutcrackers display certainly hints at such an ability. Even when a reference object such as a log or rock has been moved, the birds appear to search in a particular spatial relationship to the object.

Studies of baboons have directly demonstrated an ability to mentally rotate images, a prerequisite for recognizing an object from a novel vantage point and for maintaining correct spatial relationships as perspective shifts (for example, recognizing that the path that veers off to the left of the rock when viewed from one direction is the same path even when one approaches it from a different direction). In these studies, baboons were run through a drill very similar to that used to measure mental rotation in humans, which we encountered in chapter 3. On a computer screen, the baboons were shown a sample image—a capital letter F, for example— then two comparison images, one a plain rotated F, the other an equally rotated mirror-image F. The monkeys then had to use a joystick to move a cursor onto the plain rotated image. Two of six baboons never seemed to get the idea, but the others did significantly better than chance, scoring from 72 to 86 percent correct. (Three human subjects scored 87 to 90 percent correct on the same tests.)

This of course is not the same thing as mentally rotating a three-dimensional object, which is a far more complex task in which parts of the object are obscured and its apparent shape changes. But it is certainly suggestive, and everyday experience does not contradict the idea, that mammals and birds possess such a facility. Domestic animals have no trouble recognizing familiar objects (such as a feed bucket or a familiar person) from novel angles. Wild birds have no trouble approaching a bird feeder from any direction.

BEES DON'T DO IT

If a terrain feature stands out in some distinctive way—such as a tall tree in an open field—such feats of mental rotation may not be required at all to recognize a landmark from a novel perspective. Experiments with honeybees suggest that when bees do manage to take novel shortcuts across territory, this tactic of using distinctive landmarks as "beacons" is most likely the method they are employing.

This conclusion admittedly has proved something of a disappointment to bee enthusiasts (of whom there are a surprisingly large number). The remarkable ability of honeybees to communicate the location of food sources to their fellows seemed to promise great things about bee brains. When a bee locates a food source, it returns to the hive and puts on a dance in which the direction of the food is indicated by the long axis of the dancer's body and its distance by how long the bee emits sounds during a certain portion of the dance. In a classic experiment, Karl Von Frisch demonstrated in the 1960s that even when bees were forced to take a detour, flying around a steep hill on their way back to the hive, the dancers indicated the direction of the food source on its true compass bearing relative to the hive. These results led a number of scientists to conclude that the bees, having constructed a map of the nearby terrain in the course of their routine excursions, were locating the position of a new food source on this mental map, and calculating its distance and bearing from the hive with a sort of mental dividers and protractor.

A subsequent test in which bees were trained to visit sugar-water feeders seemed to offer even more proof of mental maps in bees. Bees trained to visit a feeder at site A were captured as they left the hive and transported to site B. Rather than returning to the hive, they steered a course directly from B to A, taking a shortcut across two legs of a triangle, even though they had never traveled that route before.

But the conclusion that the bees were consulting their mental maps to navigate these uncharted airways turned out to be premature. Fred Dyer, a zoologist at Michigan State University, repeated the experiment but set it up in a clever way that could discriminate between several hypotheses. Site A and Site B were separated by a 60-degree angle from the hive. While the landmarks that might guide bees from the hive to Site B (two patches of woods) were readily observable from A, the reverse was not true; Site B was located in an abandoned gravel quarry, and bees leaving B would have to fly at least 60 feet above the ground to be able to spot any of the landmarks that the bees might have followed from the hive to A:

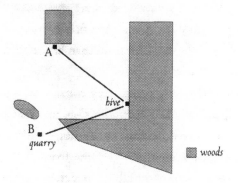

In pilot experiments, the bees never flew higher than 30 feet off the ground before they chose a direction and set off. So it was clear that bees released at B could not set their course using visual cues. Dyer noted that there were three different possible outcomes depending on how the bees were navigating. First, the bees might simply be following a learned compass bearing from the hive, without using a map or following landmarks at all. If that were the case, then the bees that were heading for A before they were captured and transported to B would be expected simply to fly off from B into the wild blue yonder at that same preset compass bearing they would normally follow from the hive to A. A second possibility is that the bees possess true maps; in that case they should have no trouble calculating the geometrical relationship between B and A, and should find it equally simple to find the shortcut from B to A as to find the shortcut from A to B. Finally, if the bees fly shortcuts using visual landmarks, they ought to find A to B much easier than B to A.

The results were very distinctive: Bees trained to feed at B that were captured and released at A flew almost straight toward B. But bees trained to feed at A that were released at B mostly flew off in the preset, hive-to-A compass direction; a minority headed home for the hive. Other experiments suggest that this latter bunch of A bees had almost certainly visited the area of Site B while feeding on wild floral sources of nectar and thus knew the way back to the hive from there. When bees were trained to visit the feeder at B and then were retrained to visit the feeder at A, the same experiment resulted in nearly all the bees heading straight home when released at B. Finally, to make sure that there was nothing fundamentally difficult about the short cut, Dyer trained bees to fly from B to A and A to B; bees which had had this experience of traveling the shortcut route then had no difficulty with the displacement experiment.

So Dyer's findings were completely consistent with the hypothesis that when bees do take shortcuts, they use visual landmarks as their guide. When those landmarks are not visible, the bees return home if they already know the way home from previous experience, or just continue on the now totally incorrect, preset compass bearing if they do not. The results appear to completely rule out the map hypothesis, for if bees that had been trained to visit both sites cannot plot their geometrical relationship and figure out a shortcut, no bees can.

So much for shortcuts, then. But what of the bees' dances? The dance obviously isn't some game of charades acting out the landmarks the bee saw along the way to the food source. (One word, sounds like bee.) The bees are able to indicate distance and direction with impressive reliability, and if they aren't establishing these parameters by reference to a mental map, what are they doing?

Bee researchers seem to love devising fiendish tricks to play on their subjects in order to answer such questions, and there are few more fiendish than the one that Wolfgang Kirchner and Ulrich Braun devised. First, they glued tiny magnets (weighing only 7 milligrams apiece) to the backs of honeybees trained to forage at a feeder about 30 feet south of their hive. Then, as the bees approached the feeder, the experimenters snared them with a small magnetic wand, held them tethered in a 3-foot-long wind tunnel set up at right angles to their hive-to-feeder flight path, and forced them to fly due east for 30 to 90 seconds. At the speed of the wind produced by the wind-tunnel fan, that flight time corresponded to a distance of about 500 to 1,500 feet. When their flight was completed, the bees were released at a second feeder set up at the far end of the tunnel. The return trip was then carried out in reverse: after flying through the tunnel due west for the same duration as on the outward voyage, the bees were then released at the first feeder, and so back due north to the hive under their own power.

If bees establish and report the location of a new food source by pinpointing it with reference to surroundings previously surveyed and incorporated into an established map, then these bees should not be fooled by their artificially extended trip. But they were: The dances of these bees reported a food source almost due west, rather than due south, and at a distance not of 30 feet but of hundreds to a thousand feet way.

The bees, in other words, measured distance and bearing through some form of dead reckoning, and that is no mean feat. Their brains in effect kept track of the length of their flight and the turns they took along the way. Other insects have been shown to have this ability too, and it may be based on some measure of energy expenditure or of travel duration or perhaps some visual cues that indicate relative distance traveled. But the one thing the bees definitely did not do was to use landmarks (including trees and buildings) that were abundantly available in the test environment and which were certainly familiar to the bees, since the feeders and wind tunnel were located so close to the hive. The bees also clearly were not constructing or drawing upon cognitive maps.

To summarize: Bees appear to use dead reckoning to establish the distance and bearing of new food sources, which they communicate with their dances; they appear to use visual landmarks to fly established routes and to manage the occasional feat of novel short cuts. They appear to use cognitive maps not at all.

TRAVELING SALESMEN

Dead reckoning, pilotage, and homing in on visual beacons can explain a good many of the accomplishments of navigating animals. One class of experiments that is interesting—and easy enough and rather fun for anyone who owns a dog or a horse to try for themselves—is to try to figure out how exactly animals do find short cuts between two baited sites. The basic technique is to start at a fixed point, lead the animal along with you as you walk in a straight line some distance, let him watch you place some food on the ground, then lead him back to the starting point. Then, aiming off in a new direction, say 30 degrees from the first line, again lead the animal out in a straight line, bait a second site, and return again to the starting point. Then let the animal go and see what he does. In published studies, dogs would almost always run straight to the first site, then cut almost directly across on a beeline to the second site, without returning to the start and retracing their steps. Dead reckoning seems the best explanation, though it is not always easy to rule out visual cues; one interesting test might be to see whether the animal's accuracy suffers at longer distances and wider angles. Another variation is to place distinctive visual markers at the bait sites and see how that affects performance. When I tried the basic experiment with my horse he didn't always do terribly well; but when he did, it was striking how unerringly he oriented himself straight toward the second site. He would finish eating the food at the first site, lift his head up, turn precisely toward the second site, and head off in a perfectly straight line. Only when he got very close to the target would he appear to search (by sight or smell) with his head down.

It bears repeating that though dead reckoning or reliance on visual landmarks falls short of true cognitive mapping, these are nonetheless quite sophisticated cognitive feats that require complex mental representations. All involve an ability to store, process, and recall information in a way that is far from trivial.

The hunt for more sophisticated spatial reasoning that more closely approximates a true map-making and map-reading skill leads through some ambiguous territory. Some of the most impressive performances have been turned in by chimpanzees. In one now-famous experiment, a chimpanzee was carried piggyback around a one-acre enclosure where it lived with several other chimps; a second person walked along and hid a piece of food in each of eighteen randomly selected spots while the chimp watched. The observer chimp was then placed back with its group and two minutes later the chimps were all let back into the enclosure. The chimp that got to see where the food was hidden not only did a remarkably good job of remembering each location and retrieving the food, but also did one better: it went from spot to spot in a different order from the one followed on its piggyback ride. And the routes the chimp chose were significantly more efficient—in terms of the total distance traveled—than chance would predict.

Finding the most efficient order in which to visit a series of sites is known as the traveling salesman problem in mathematics, and its solution is quite complex. The number of alternatives grows as the factorial of the number of sites to be visited; there are more than 6 quadrillion different ways to visit eighteen sites. The chimp, unsurprisingly, did not hit on the one optimal route out of those 6 quadrillion options, but did awfully well.

Again the question arises whether this was evidence of an animal possessing a cognitive map. These studies have frequently been cited as proof that they do. And what the chimpanzees did in remembering so many sites was unquestionably a remarkable feat of spatial memory. But at least one critic has suggested that the chimps in such experiments could have headed toward landmarks that they had noted during the piggyback phase and subsequently recognized; upon arriving in the general vicinity of each cache site, they could have found the precise spot of the hidden food by recognizing other, local landmarks (much as Clark's nutcrackers apparently do). The enclosure where the experiment was carried out was an area the chimps were thoroughly familiar with, and all landmarks were visible from all parts of the enclosure.

On the other hand, the fact that the chimpanzees' routes were more efficient than chance—and that they did not merely follow a strategy of traveling from one site to the next by choosing the closest one, but rather appeared to strike out in the direction of clusters of sites that could be scooped up with greater efficiency—does strongly suggest that the chimpanzees had mastered a sense of the overall spatial relationships of the eighteen locations.

Similar experiments with rhesus and vervet monkeys found that these animals could never remember more than six locations where food was hidden. But they, too, appeared to make route decisions that increased the efficiency of their trip—hitting more sites in less total distance traveled. In the vervet monkey test, each monkey was carried around its thirty-by-thirty-foot outdoor enclosure and allowed to see where grapes were hidden in six of twenty-five possible holes. With only six sites to visit, a strategy of always moving to the closest, next site yields an optimal route in most cases. But the monkeys seemed to show that they were thinking more than just one site ahead. In one experiment the researchers arranged things so that the monkey would start at a point equally distant from two baited holes. One choice, though, led the monkey initially to a side of the cage where only two baited sites lay close together; the other choice led to a side where four sites lay in close proximity. With unfailing consistency, the monkeys always headed toward the richer side first:

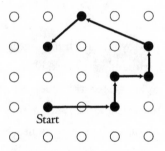

Finally, a study of wild chimpanzees in Taï National Park in the Ivory Coast also appeared to show an ability to remember multiple locations and to calculate the least distance from novel positions. These chimpanzees habitually use heavy stones as hammers to crack nuts. The hammer stones typically weigh from a pound to as much as forty pounds. Stones are rare in the forest where the chimpanzees live, and the animals frequently transport them from tree to tree as needed. By marking every nut tree and every stone within a 2-square-mile study region, the researchers were able to track the transports that occurred. As the chimpanzees go from tree to tree, they have to decide which of several nearby stones they should go and fetch. In many cases they not only appear to remember where the stones were last left, but also are able to pick the closest one, even when all of the stones to choose from are out of visual range. Out of the seventy-six cases in which the available stones were all out of sight of the tree where a chimpanzee was foraging (more than about 60 feet away in this forest)—and in which the researchers were able to identify all of those possible options—the chimps chose to transport the closest stone in forty-eight cases, or about 63 percent of the time. The authors of this study, Christophe and Hedwige Boesch, note that if the chimpanzees were merely searching randomly for stones within a 1,000-foot radius of a foraging tree, then the probability of their picking the closest stone averages only about 26 percent.

There are still nagging doubts about what exactly these chimpanzees are doing. A random search that combs the immediate vicinity first, before expanding to a greater radius, would always lead to selecting the closest stone first, for instance.

PLACE CELLS

Evidence that both rats and human toddlers can orient themselves by the overall geometric shape of their local environment probably comes the closest to establishing the existence of a true cognitive map in animals. Tiny electrodes inserted into a rat's brain—specifically, the part of the brain known as the hippocampus—have revealed that the brain makes an extraordinary, automatic calculation of spatial position: Different cells in the hippocampus fire when the rat is in different parts of a room. These so-called place cells map out the entire space of the local environment, overlapping slightly with one another. The geometric arrangement of the place cells within the hippocampus does not match the geometric arrangement of the real-world places they correspond to in any obvious fashion. In other words, it is a strange sort of map; it is as if we cut out all the countries in the world, mixed them up, and glued them back down in a new order. But however the brain physically arranges these place cells, the correspondence between a given place cell and a given place is complete and reproducible. Whenever the rat is in one corner of a rectangular box, one particular place cell fires; when it is in the middle, another particular cell fires. Within a few minutes of entering a new environment the hippocampus has constructed a new map.

How is the brain making this calculation? That is, how does the rat's hippocampus "know" where the rat is? Additional studies have shown that rats can put together reliable place-cell maps even in total darkness and when smells and sounds that might help distinguish one side or part of the room from another have been scrupulously eliminated. That strongly suggests that the animals are using some form of dead reckoning from walls or corners to establish where they are, and so to build up a mental map in the hippocampus.

The usual sort of playing-devilish-tricks experiments confirm this. When rats have learned their way around a small room what happens if, unbeknownst to them, the dimensions of the room are stretched? In effect, the place cells stretch, too. One place cell in the original room might correspond to the center of the room, which is, say, ten paces in from each wall. But in the stretched box, that same place cell now covers two locations: If the rat is moving from one wall toward the center, it now fires not in the center, but ten paces in, as before. When the rat is approaching the center from the opposite wall, it also fires ten paces in.

This map-making and map-reading system is clearly automatic; it is a matter of the extremely clever way the hippocampal place cells are wired up in the first place. It is a system that is extremely sensitive to overall geometry of a local environment; in practice it allows an animal to know its relative position within a familiar area without even knowing it, so to speak.

A clinching piece of evidence that this map system is fundamentally geometric, rather than based on visual landmarks, comes from some clever experiments with rats and with human toddlers. In both cases the subjects were placed in a rectangular room and allowed to watch as a desirable object was concealed in one corner (rat food or a toy, as the case may be). The subjects were then deliberately disoriented, either by being placed in a second, identical room (rats) or being gently turned round in circles with their eyes closed (children). A rectangle has four corners, of course, and if all you can do is recognize corners—but not the overall geometry of the room—you would be likely to search all four corners randomly in search of the object after being disoriented. But if you do have a grasp of the geometrical properties of the room, you can eliminate two corners right off the bat, because one pair of diagonally opposite corners does not look the same as the other, no matter how disoriented you are. (Draw a rectangle and mark the upper left and lower right corners with an X. Now turn the rectangle upside down. The rectangle looks exactly the same. You can't tell which X is which. But even if you hadn't marked Xs on them, you could still tell those corners from the other diagonal pair; they are geometrically distinct. In both the right-side-up and upside-down rectangles, you can find the "X" corners by facing a long wall and choosing the corner all the way to your left.)

Disoriented rats and children consistently search for the reward in the two corners that are geometrically similar to the one where they saw the reward concealed originally; they search half and half in each of the two locations. Thus they clearly have a spatial sense of the room's over-all geometry. Even more interesting is what happens when one corner or side of the room is given a distinct visual marking. Although adult humans immediately use the visual cue to figure out which end is which when they've been disoriented, rats and eighteen-to-twenty-four-month-old children do not. Reorientation is accomplished solely on the basis of shape of the local environment.

Having a map is only a part of what is needed to navigate, though; the other requirement is a compass that allows an animal to calculate its current heading within that map framework. Neurophysiological studies have shown that the rat has cells in another part of the hippocampus that appear to fire according to the direction it is facing within the environmental frame, regardless of where the rat is at the time. Significantly, when rats are disoriented in experiments such as the one we just saw and their map gets turned upside-down, this internal map-compass turns with it.

Sometimes rats *can* get their map and compass turned the right way by using a visual cue. For example, rats are placed in a circular room with a white card at one spot on the wall and allowed to find their way around. Then the lights are turned out, the card moved 90 degrees and the lights switched back on. The rats often will rotate their place-cell fields to re-align with the cue card. But not always. Visual cues seem to form a secondary mechanism for orienting a map. But it is not hard to see how the hippocampal map can be used to form associations with visual landmarks or memories of important things or events. The hippocampal place cells fire automatically when the animal is in the corresponding location in space; if that signal coincides with a visual input of a distinctive landmark or an important experience (a tree full of nuts or the sight of a hammer stone if you're a chimpanzee; a bird flying out of the woods and scaring you if you're a horse), then new objects can be added to the map—much as a tourist map might include the location of motels or historic sites.

Having a map, and a compass to go with it, are the prerequisites for true navigation, and such a combination of hard-wired devices in the hippocampus could explain some of the more dramatic feats of animal orienteering. Pigeons are probably the most remarkable animal navigators, for they have clearly shown an ability to set an accurate course for home even from places they have never before visited and even when transported there blindfolded and disoriented. Their compass is reasonably well understood: Pigeons are able to use a comparison of the position of the sun with their internal clocks to determine compass direction. In one experiment, homing pigeons were placed for several days in a closed room where artificial lighting was switched on and off six hours out of phase from actual sunrise and sunset. When these clock-shifted pigeons were released in early morning, they headed not north to home but almost due west. (Control pigeons that had not been clock-shifted headed north.) Pigeons may also be able use the earth's magnetic field as a back-up compass on cloudy days.

How they construct their maps, though, remains a mystery. One theory is that the pigeons while at home learn to associate specific odors with the wind direction at the time that they sense them. In other words, they construct a smell map; when released at a new location, the local smell can tell their direction relative to home.

PICKING, POKING, AND SMASHING

The ability to make and use tools is almost the defining fact of modern human existence. The accomplishments that we as humans prize the most are often our feats of technological mastery. So it is natural for us to be impressed by tool use in other animals. Tool use has often been taken as a critical sign of intelligence; Donald Griffin argues that it strongly suggests underlying, conscious thinking about something the animal is trying to accomplish.

Yet there does not seem to be a lot of correlation between tool use and other behaviors that imply cognitive sophistication. Many insects use tools, but few birds and mammals do. Capuchin monkeys regularly employ an assortment of tools in the wild, including sticks to extract food, rocks as hammers to crack nuts, and a variety of objects as missiles to drop on snakes or scare away other potential predators. But gorillas and orangutans, members of the great ape family that is the closest to humans, make little or no use of tools in the wild. Sea otters and ground squirrels exhibit tool use, while dolphins and porpoises and whales do not.

One source of this apparent contradiction in tool use patterns is our definition of what a "tool" is. Scientific journals are full of hair-splitting arguments over exactly what qualifies as tool use (is it tool use when an elephant scratches its back against a tree, or only when the elephant picks up a branch and uses that as a scratcher? Is it tool use when an Egyptian vulture drops an egg on a rock, or only when it drops the rock on the egg?). But all of these definitions still have the effect of lumping together many fundamentally different kinds of behaviors, which probably demand very different levels of cognitive sophistication to explain. We tend to throw them together because they all resemble superficially what we think of us as tools. The inevitable result is to exaggerate the significance of some behaviors while failing to make some basic distinctions about what an animal is actually up to.

The use of "tools" by insects, for instance, follows highly stereotyped patterns—prime illustrations of the brainless intelligence of evolution. The solitary wasp holds a small pebble or piece of wood in its jaws and tamps down the soil around the entrance to its burrow where it has laid an egg. This apparently makes the burrow less visible and harder to break into—important defenses against a number of species of parasitic flies and wasps that search for wasp burrows to lay *their* eggs in the bodies of the developing larvae. But to concentrate on the wasp's act of using a "tool" is to miss the larger story. More remarkable than its use of a pebble, after all, is that it "knows" to conceal its entrance in the first place. More remarkable still is the wasp's deliciously gruesome stratagem for feeding the developing larvae: it paralyzes its prey of caterpillars or crickets or other insects, carries the still-alive but immobile prey to the burrow, and drags it underground where the emerging larvae can consume it. All of these actions could well be viewed as far more "intelligent" than its use of a tool. Yet all have been shown to be the result of genetically programmed rules that the wasp applies with singular stupidity. After laying an egg, the female will return several days later with additional prey; on her third visit the female may bring a supply of five or six caterpillars, also conveniently paralyzed to retard spoilage, and seal up the nest and leave it for good. Some famous nineteenth-century experiments with solitary wasps showed how the female's procedure of delivering prey is followed with utter literal-mindedness. The female puts the prey down right next to the burrow entrance, digs open the tunnel, enters and inspects the burrow, then returns to the top and pulls the prey down inside. The French entomologist Henri Fabre found that if he moved the prey a little bit away from the tunnel opening while the female was on its inspection tour, it would not simply pull the prey down upon coming to the surface, but would repeat the entire procedure: It would retrieve the prey and place it next to the tunnel opening, enter and inspect the tunnel, then come up to pull the prey down again. Fabre repeated his experiment over and over; each time when the wasp returned to the surface and found the prey out of position, it would start from square one again. The wasp could be made to take forty or more inspection trips in this fashion. It is hard to view pounding the entrance closed with a pebble as a product of creative intelligence in light of this evidence.

Insects have been reported to use tools in many other situations, but all are equally stereotyped. Some species of ants use bits of leaves or mud as sponges; they soak up liquid foods and then carry the saturated sponges back to the nest. Some assassin bugs dangle a dead termite as a decoy to lure others out of the nest, which they then capture and eat. Much of what birds and nonprimate mammals do with tools is equally ritualized. A number of species of birds will use sticks to pry insects out from a crevice. Although some degree of learning is clearly involved as the birds gain experience in using these probes to best effect, the basic instinct appears to be there from birth, and is really just an extension of the basic instinct to probe directly with the beak. Galapagos woodpecker finches use spines and twigs to probe for insects in the bark of trees; when a fledgling woodpecker finch was taken from its nest, and raised in isolation, it showed an interest in playing with twigs from an early age. When presented with an insect in a hole, the bird would at first drop the twig and try to dig out the bug with its beak, though over time it began to use the stick as a probe.

Nest-building is another elaborate but thoroughly stereotyped behavior in birds. Even the most elaborate nests, such as the woven, roofed bowers of weaverbirds, are constructed through a rigidly determined sequence of motions. Does a bird have a mental image of the nest he wants to build? He need not; each stage in the sequence appears to be triggered by the completion of the previous stage, and if a nest needs repairs the bird will simply pick up the sequence at the corresponding point. The motions appear to be wired in firmly in a bird's brain: Weaverbirds reared in isolation are able to construct typical nests when supplied with the right grasses. Even males deprived of the right building stuff in captivity will go through the same motions, and attempt to weave a nest from their own feathers. Canaries or pigeons deprived of nesting materials altogether will go through the motions of making an imaginary nest, or will pick up a pencil or other similar object that might be available and repeatedly manipulate it to build a nonexistent nest.

Weaverbirds do get better with practice, but this appears to be mainly a result of learning which sorts of grasses are most easily woven and how to tear tall grasses into long, neat strips. This is hardly different in kind from the sort of learning that takes place in the course of development in most birds and mammals that have to practice in order to perfect such skills as hunting, flying, and recognizing dominance hierarchies.

PLANNING AHEAD

Higher primates tend to show much more flexible and creative use of tools. Studies of wild chimpanzees have shown regional differences in tool use among different bands of chimps, which in itself suggests that these behaviors are more than merely stereotyped instincts. And chimpanzees are apparently unique in their regular practice of preparing and modifying tools to suit their intended purpose. They will cut sticks or grasses to a specific length, strip off leaves or bark, and in the case of chimps in the Taï National Park, sharpen the ends of sticks with their teeth. The Taï chimpanzees make two general kinds of sticks, shorter and thinner for digging out marrow from bones and picking out bits of nut kernels from the remains of crushed nutshells; longer and fatter for fishing out ants or honey from insect nests. These differences were clearly appropriate to the task, as the holes to reach ants or honey are indeed larger and deeper. In only about 6 percent of observations did the chimpanzees make on-site modifications of the tools they had created ahead of time—suggesting that they were able to make the right tool for the job in advance.

Chimpanzees in the wild also will use sticks to probe for bees at the entrance to a hive before breaking it open to eat the grubs and honey. Adult bees react to an intrusion by blocking the entrance with their abdomens; when that happens the chimpanzees use their stick to disable the bee. As we have already seen, chimpanzees also use stones as hammers to crack nuts. The chimpanzees at Gombe Stream and Mahale Mountains National Parks in Tanzania use leaves or sticks to clean dirt off themselves, and to wipe off semen after copulation, though the Taï chimps have not been observed to do so.

The ability to use tools in this fashion—to prepare them ahead of time, to transport them some distance, and to retrieve a necessary tool when needed—strongly suggests an ability to maintain a sophisticated representation of both objects and their use. In laboratory experiments, there is a reasonably consistent pattern in that great apes are better able to solve problems involving tool use than are monkeys. Even though orangutans scarcely use tools in the wild and capuchin monkeys make extensive use of tools under natural conditions, orangutans (like their great-ape cousins the chimpanzees) showed a substantially greater ability to solve a task requiring the use of sticks to poke a piece of food out of a clear tube. When the sticks were bundled together with a rubber band, making the entire bunch too fat to insert into the tube, the apes in the study immediately unwrapped the bundle and pulled out an individual stick. The monkeys tried to insert the entire bundle. When a misshapen stick was provided, the apes tried to straighten it out before using it; the monkeys did not.

Most people would not hesitate to conclude from such evidence that apes are smarter than monkeys. But that may be jumping to a conclusion. Much of the tool use exhibited even by chimpanzees reflects an extension of their natural manipulative abilities. Overall, tool use per se may be more a natural consequence of an animal's abilities to hold and use objects—and a reflection of ecological necessity—than a manifestation of brain power per se. The capuchin monkeys are a perfect example. They readily use tools and always solve complex tool-use problems in the lab—even as the extravagant mistakes they make reveal a total lack of understanding of the spatial relationships involved. In the wild, capuchin monkeys have been reported to use sticks to kill a snake, as weapons against other monkeys, and as probes for food, much like chimpanzees. But some of the errors they have made in lab tests were comical. When presented with a peanut in a Plexiglas tube and a variety of sticks, they showed little ability to reason insightfully about how to solve the problem. When the monkeys were given several short sticks, which could push the peanut out if inserted one after another from one end of the tube, one monkey tried putting one stick in one end and another stick in the other end. When given the bundle of sticks held together with a rubber band, one monkey tried putting the rubber band in the tube, even after a previous successful technique had been used. "The monkeys did not know why a successful technique was more appropriate than unsuccessful attempts," noted researcher Elisabetta Visalberghi.

Studying how an animal uses tools clearly says much more about its cognitive abilities than studying whether an animal uses tools. *Whether* an animal uses tools or not may be largely a matter of an animal's ecological niche and anatomical capabilities. Capuchin monkeys naturally specialize in what has been called "destructive foraging." That is, in their natural habitat they "manipulate, break apart, and bang anything at hand," says Visalberghi. That allows them to obtain food that other competing species cannot.

Likewise, the use of tools by other non-great-ape animals also seems to be closely tied to their natural manipulative habits and the evolutionary adaptations they have made to their ecological niche. Elephants constantly use their amazingly strong and versatile trunks to manipulate objects in the course of feeding—bushes, branches, grasses, even entire trees. The overwhelming majority of instances of tool use observed in wild and captive elephants consists of swatting flies or scratching using vegetation held in the trunk. A bird's beak is likewise a remarkable and versatile tool for exploration, digging, preening, nest building; when birds use tools it is invariably with a stick or such held in the beak and used as an extension of the beak. And at least some of the versatility that chimpanzees show in tool use compared to other mammals directly reflects the anatomy of the chimpanzee's hand. Like humans, chimpanzees can use both a "power grip"—squeezing an object—and a "precision grip" between thumb and forefinger which allows more delicate manipulation of an object. Animals use tools, in other words, because they can. This is Euan Macphail's point, that if dogs had hands we would have no reason to think they wouldn't use them.

Interestingly, there is some evidence that rather than intelligence being the thing that makes tool use possible, tool use may be the thing that creates intelligence—intelligence which can then be exploited for other ends. Many cognitive abilities may in fact piggyback on motor skills and the associated mental wiring needed to carry them out. (The close relationship between spatial memory and locomotion is one example of this; the very act of moving around appears to play an important part in the ability to compute spatial locations.)

The need for circuitry that can control the human hand's precision manipulations may in a very odd way have even set the stage for language development. Humans possess a unique manipulative ability to throw objects accurately. While chimpanzees do throw things as a threat against an intruder, most accounts of accurate throwing by chimpanzees are, as neurophysiologist William Calvin has pointed out, anecdotes that record their occasional—and quite possibly lucky—

hits. No one has recorded how many misses they make. ("I think we can assume that if chimpanzees were accurate enough to bring down game with any regularity, we would have heard about it," Calvin says.) Any limb movements that last less than about an eighth of a second leave no time for midcourse corrections: the nerve cells just don't fire fast enough to allow for feedback adjustments. That means that an action like hammering, or throwing a baseball, requires that dozens of muscle movements be precisely planned out by the brain in advance, a process that Calvin compares to punching a piano roll ahead of time and then letting it play automatically when it's show time. The problem of throwing a baseball accurately is further compounded by the fact that the ball has to be released with incredibly precise timing as the arm swings forward. To hit a rabbit-sized target 12 feet away, the "launch window" is only about a hundredth of a second wide. Release the ball outside of the window and you'll miss. To hit the same target at double the distance reduces the launch window to a thousandth of a second. And according to fundamental statistical calculations, each doubling of distance would require sixty-four times as many neurons to manage the calculation.

What is the connection to language? The use of language, like throwing a baseball, involves complex sequencing of neural events. Both the muscle movements in speaking and the formulation and comprehension of words require rapid construction and decoding of sequential patterns. So too do such uniquely human (and otherwise inexplicable, in evolutionary terms) inventions as music and dance. It is intriguing that two very different brain disorders are both related to injury to one region of the left cerebral cortex that is activated during facial movements and in listening to speech. In aphasia ("loss of speech"), patients lack the ability to understand and formulate linguistic symbols. They have trouble finding the right word; they may substitute one sound for another; they have trouble reading, writing, speaking, and understanding. Patients suffering from apraxia ("loss of action") lack the ability to carry out purposeful movements of the hand and arm. This could merely be a coincidence of anatomy; two separate pieces of the brain, one that controls speech and one that controls arm movement, may just happen to be located near one another.

But it might also be no coincidence at all. The wiring that first evolved to allow the sequencing of complex arm movements could more or less accidentally have bequeathed us the glorious gifts of dance, and music, and language.

6

SPEAK!

THE BELDING'S GROUND SQUIRREL IS A GREGARIOUS
creature that lives in open country with little forest cover. It digs burrows, but spends a lot of its day in the open on rocks and logs, and so
is quite vulnerable both to aerial attack by hawks and other raptors and
to ground attack by mammalian predators such as badgers.

The two types of predators use very different offensive strategies.
Hawks rely on speed, badgers on stealth. And the ground squirrels,
quite naturally, rely on two very different defense strategies. A squirrel
that spots a badger will retreat to the opening of its burrow, maintain
an erect, vigilant posture, and monitor the situation. This is a widespread strategy used by prey animals. It in effect lets a predator know
that its attempt to sneak up has failed, and that the predator might as
well give up and go home. The squirrel is saying (in the compulsory anthropomorphic language of evolutionary selection), "Look, we could
both waste a lot of time what with you chasing me and me running
away or escaping down my burrow, but let's face it—now that I've seen
you you're not going to win that race. Your only hope of catching me
was if I didn't notice you first. The fact that I'm standing up here so
conspicuously, staring right at you, ought to convince even a walking
shaving brush like you that I'm on to your little game."

If a ground squirrel spots a hawk, which can dive at incredible speeds, it has no time for such extended chats; it runs for the nearest cover of brush it can find and hopes for the best.

Researchers have discovered that ground squirrels also make two very different sounds depending on the type of predator they spot. The "hawk alarm" is a high-pitched whistle of a single, pure tone. The "mammal alarm" is a rough chattering.

This would seem to be strong evidence in support of the semantic view of animal communication that has long colored both popular and scientific accounts. The idea that animal sounds amount to communication seems so obvious a notion that it has long gone unquestioned. Human language is manifestly about a mutual transfer of information by a cooperating sender and receiver using a commonly understood code. So, too, animals have long been viewed as having "food calls" or "mating calls" or "alarm calls" with clear semantic meanings. The animals' vocabulary may be sharply limited, but when Tarzan says *"Ka-goda!"* (which, we are reliably informed, means, "Do you surrender?" in the ape language) or when a ground squirrel says *seeet* ("Hawk!"), they are using words just as we do. And clearly other ground squirrels react as if they understand what ground-squirrel words mean. Hearing a "hawk" call will send squirrels running as fast as they can for cover; hearing a "badger" call will send them to their burrows, where they stand erect and join the badger-monitoring force.

VENTRILOQUISTS AND COWARDS

But there is something missing in this explanation—namely an explanation. What the ground squirrel *does* in the face of a predator shows it is out to save its own skin, a perfectly respectable strategy in the game of natural selection. Self-sacrificing heroes are rare in the animal world for the simple reason that they do not hang around to pass on those noble traits to future generations. Animals that produce relatively few offspring will of course take risks to protect their young, but that's different—they have already passed their genes on to those young, which thus represent an investment. To put it in less anthropomorphic and teleological language: Which gene is more likely to survive in a popula-

tion over time? One that codes for an instinct of parental protective-
ness, or one that does not? The reason most of us have such strong instincts of this kind is that we inherited them from our parents who, by
virtue of their having this trait themselves, made sure that we survived
long enough that we might become parents ourselves.

But how, we ought equally to ask, does shouting "hawk!" or
"badger!" help a ground squirrel save its own skin? Perhaps, by help-
ing the entire group or species survive? That is a nice idea, but one
that for the most part turns out to be mathematically impossible by
the rules of natural selection. The fact is, natural selection doesn't
care if a species survives. It is individuals that have genes, and generally speaking a gene that makes an individual act like a Boy Scout is
evolutionarily unstable. Selection ought to favor squirrels that, while
everyone else is offering selfless assistance, keep mum when they spot
an imminent attack. Tell the world about it and everyone is equally
on alert. Making a sound could also conceivably attract an attack to
the caller. But sneak off quietly and leave the suckers to be the ones
eaten by a hawk, and you come out ahead. So it would seem that the
ideal strategy is to be the one noncaller in a community of callers.
But if that's the case, how could the instinct to call out have arisen in
the first place? As soon as a population began to contain some callers,
the noncallers would beat them every time. The survivors—who
would survive to pass on their genes to the next generation—would
be the sneaks who kept quiet.

The answer only makes sense if we drop the idea that the squirrels'
calls are semantic utterances, vocabulary words in some language. The
first interesting clue comes from the acoustic properties of the squirrels' calls themselves. The source of a high-pitched, pure tone is very
hard for mammals or birds to locate in space. So for starters, a squirrel
making a "hawk" call is not giving away its location to the hawk. Some
experiments have found that such high-pitched pure tones are not only
difficult to locate but actually confusing. That is, they don't merely jam
the hawk's location-detection apparatus but spoof it. When hawks
were played tapes of various sounds at various pitches, they generally
turned their heads very accurately toward the concealed source. But

when played a robin's high-pitched *seeet* call—which the robin, just like the ground squirrel, uses as an alarm call when it spies a hawk—the hawks turned their heads 90 degrees the wrong way.

The other even more significant point is what a ground squirrel's *seeet* accomplishes in the way of manipulating the behavior of other ground squirrels. It's good to have the other suckers left in the open while you run for cover, but in fact if your aim is not to be eaten yourself, it's even better to create a pandemonium where everyone's running for cover. Being the only one to run is a way to make yourself highly conspicuous to a hawk. So there is a strong selective advantage for a squirrel that spots a hawk to call out—because by so doing he reduces the chances of getting killed personally. Likewise, a "receiver" who runs for cover when he hears such a signal will come out ahead personally compared to one that ignores it.

Note that it is the mutual exploitation of two self-interested parties that makes this happen. Each is acting in its own self-interest in a way that reinforces the behavior of the other—the sender to send and the receiver to receive. The sender is "manipulating" the action of others; the receiver is "assessing" the action of another and responding. The net result is an "informative" signal. But this phenomenon didn't evolve because communication and information sharing is a good thing, or even a goal. It evolved because ground squirrels that make a shrill whistle when they see a hawk tend to survive and pass on this trait, and ground squirrels that run for cover whenever they hear a shrill whistle tend to survive and pass on this trait. The sender and receiver don't even have to be aware of the existence of one another for such evolution to take place.

This "management/assessment" approach to understanding animal signals is the work of Donald Owings, a psychologist at the University of California–Davis, and Eugene Morton, an ornithologist at the Smithsonian Institution's National Zoological Park. Their insight differs sharply from the traditional "informational" perspective in many ways, but most notably in focusing on the acoustic properties of the signals animals use. In human speech, sounds are of course arbitrary. It doesn't "mean" anything that the word *elephantiasis* is long or has a hissing sound to it at the end. But animal sounds are not arbitrary, and the very fact that certain sounds serve certain functions consistently across many species reveals an evolutionary purpose that is hidden to those who insist on looking at animal sound through the human experience of the semantics of information flow.

Robins and chickadees, for example, have alarm calls very similar to those of the ground squirrel. A robin will make a *seeet* call, a chickadee a *seee*, when it spots a hawk. And apparently for much the same reason—to avoid giving away its own location and to prompt action by others that betters its own odds of survival.

A chickadee will also makes a low, somewhat harsh *chickadee* when it spots a cat, much as the ground squirrel chatters at a badger. Here the manipulative goal is to call attention to the intruder; the more squirrels or birds that are on guard, conspicuously monitoring its presence, the safer the caller is. An extreme form of this behavior is "mobbing," where birds will join together and harass a predator with loud calls and even aggressive feints. It is not uncommon to see blue jays swooping down on a cat, or sparrows surrounding a hawk in mid-air. Again the goal is exactly the same: to take away the predator-by-stealth's element of surprise so that it gives up and goes home.

It is no coincidence that the low, harsh chatter call is very easy to localize—it is again a call acoustically well designed for its purpose. A receiver can readily tell where the action is, the better to help monitor the intruder's progress. Again, there is no semantics at work here. To label this call a "mammalian intruder alert" or a "word" that *means* "mammalian predator!" is to miss the evolutionary story behind it.

ANIMAL ESPERANTO

One strong prediction of the management/assessment approach is that some features of animal communication will cut across many species. In other words, instead of pursuing the meaning of "words" in Old Middle Low Lion and Basic Blue Jay, we ought to look for common physical features in the sound patterns themselves that many different species use. Pursuant to the traditional semantic or informational ap-

proach, animal communication studies have generally set out to catalogue every sound an animal makes, simultaneously noting what the animal was doing at the time. Then various labels are attached to the
various sounds—food calls, mating calls, and so on. That very approach, of course, assumes as a given that the purpose of animal signals
is communication; that the animals have in place a system that evolved
with no other purpose than to permit the transfer of messages, a sort of
universal telegraph that can carry information without any special regard to its content; and that our main challenge is so to decode the semantic meaning of those messages in each species.

One problem that has consistently dogged these efforts is that the
same "word" is used under highly varying circumstances. By the same
token, different signals, even different regional "dialects," are often employed in identical situations that ought to be assigned identical
"meanings" under this methodology.

Perhaps the most universal characteristic of animal signals is their
adherence to a basic rule of pitch. High-pitched tones (a dog's whine)
convey appeasement, fear, an overall nonthreatening state. Deep, rough
sounds (a dog's growl) convey aggression, hostility, threat. Morton analyzed the sounds of fifty-six bird and mammal species and found this
pattern consistent throughout. Dogs growl, but so do opossums, rats,
African elephants, pelicans, ring-necked pheasants, and Carolina chickadees; dogs whine, but so do guinea pigs, wombats, rhinoceroses, bob-
whites, barn swallows, and mallards. And so, in fact, do people. If you
talk to a baby, you will find yourself naturally using a soft, high-
pitched, tonal sound. If you are telling a dog (or a driver who has just
cut you off) to get out of your way, you will find yourself with equal instinct using a rough, low—even growling—sound. Morton calls this
pattern of relationship between an animal's motivation and the acoustic
structure of the sound it employs a "motivational-structural rule."

Now why should you—or a rhinoceros or a sandpiper—obey
these rules? The answer is a perfect illustration of how senders and receivers use one another to evolve signals without even knowing it. Big
things make low sounds. A longer string or a longer organ pipe makes a
lower sound than a short one. It is thus a physical fact that big animals

make lower sounds than small ones. Now big animals did not go around making low sounds *in order* to show that they were big. Nonetheless, animals that learned to avoid low sounds saved their hides; animals that learned not to run at every peep and squeak saved a lot of wasted time and energy running from nonexistent threats.

Back to the senders: once animals began assessing signals based on pitch, the senders could exploit that fact for manipulative ends. An animal that sought to get another out of its face could use a lower pitched sound. An animal that wanted another to come near could use a high-pitched sound. Back to the receivers: If there were no social advantage in assessing such manipulative uses of signals—signals that were once based on a true reflection of physical reality—the story might have ended right there. Receivers might simply have come to ignore such signals altogether. But there was an advantage: A growling animal *is* one that it pays to avoid—whether he is big, or whether he is just feeling mean. A whining animal is one that one need not avoid—whether he's small, or whether he's just feeling "friendly." These signals, in other words, have become "ritualized." A social function has piggybacked a ride on a physical fact. Ritualized signals furthermore serve a useful purpose by eliminating a lot of unnecessary violence among group-dwelling animals. If a growl reveals the true motivational state of a dominant animal that is ready to fight, a subordinate animal can avoid the trouble by backing off before the first blow is struck.

This evolutionary perspective underscores the important fact that signals don't evolve because they "mean" something; they evolve because they work. Again, the unwitting feedback between sender and receiver works to create signals that are informative without any conscious intent for them to be so. Sender and receiver continually exploit one other; when the exploitation is mutually reinforcing, a stable and informative signaling pattern emerges.

Morton suggests that the key question to ask is not what an animal is trying to say, but what is it trying to accomplish. Natural selection will over time favor the acts of sending/manipulating and receiving/assessing that are effective. The synergy between the two arises because the effectiveness of the sender's signal in achieving his desired end depends on the response of the receiver; the effectiveness of the receiver's response depends on the relation the signal bears to something of importance to him. The combined effect of these two pressures is to keep signals relatively "honest." A sender whose signals bear no proper relation to something a receiver cares about will quickly lose its effectiveness—the receiver will simply learn to ignore them.

From this evolutionary point of view, calling a signal "deceptive" is rather silly. The whole point of animal signals is manipulation. Most of the time selection favors signals that, from a neutral informational viewpoint, are "honest." But the only reason animals send signals at all is that it confers a selective advantage to themselves to do so. Selection selects what works. It is an unwarranted a priori assumption on our part that signals are supposed to be semantic, informative, and honest—and that when they depart from that, there is some deception (or even deliberate, calculating, Machiavellian lying) at work. Occasional signals that elicit responses disadvantageous to the receiver can be part of a stable strategy as long as they don't become so frequent that they constitute a selective force against the receiver's paying attention.

Donald Griffin, in attacking the contention that animal signals lack semantic meaning, has caricatured this view as the "GOP" position—animal sounds are but "groans of pain," uncontrollable utterances that reflect internal physiological processes. But Morton's argument suggests nothing of the kind. Evolved signaling can be intricate, effective, even "devious" or "Machiavellian"—without being either conscious or possessing semantic meaning. And like other behaviors under cognitive control, there is little doubt that in both sending signals and acting upon them, animals are often responding in a complex fashion to circumstances. This sort of unconscious decision-making is absolutely no different from many other things animals do that involve a combination of genetically programmed and learned behaviors. Griffin suggests that animal communication is a window on the conscious minds of animals—that "it seems likely that animals often experience something similar to the message they communicate."

But if the messages are not messages at all, and if indeed they are un-conscious products of natural selection, then Griffin's argument is cast in a very different light.

REGIONAL ACCENTS

The complexity of bird song has come in for special attention among those who see in animal communication a token of consciousness—or even "group consciousness," since the songs of some bird species take on regional variations and local dialects. Certainly there are some im-pressive cognitive processes at work in the species that learn their songs. Mockingbirds continue to add new songs to their repertoire throughout their lives; brown thrashers, the virtuosos of the avian world, sing up to 1,200 different songs. Though born with a template that determines the general outlines of an acceptable song pattern, songbirds need to hear both themselves and other members of their species in order to develop a typical song. When raised in silence, young songbirds develop only a very rudimentary "subsong." (It is im-portant to keep in mind, though, that not at all species which produce songs operate this way. The vocal repertoire of many birds, and nearly all mammals, is hard-wired from birth. A deaf pigeon or a chicken will develop perfectly normal pigeon or chicken sounds. A turkey raised by foster parents of a different species will develop perfectly normal turkey sounds.)

But a look at the way songs are used by birds quickly makes clear that melodic complexity and regional variation does not reflect any cor-responding semantic complexity. Many studies have confirmed that bird songs all serve one basic purpose: staking out territory and attract-ing a mate. Replace a male bird in the woods with a loudspeaker play-ing his song, and neighboring birds of his species will still keep out; replace him with silence and they quickly move in to take over his aban-doned territory. Female mockingbirds have been shown consistently to prefer males with the most elaborate songs.

Rather than a window on the animal mind, bird song appears to be a record of an evolutionary arms race. As we have seen, receivers and

senders generally have competing interests. Sometimes, those competing interests act synergetically to produce "honest" signals. But sometimes the evolutionary pressures lead in a different direction, to a
continual escalation of tactics on both sides.

The basic fact about bird song is that it is a signal used for long-
range communication with members of one's own species. It is an
acoustical fact that even sounds well suited to long-range communication become distorted and degraded as they travel through space. They
lose amplitude—they get softer, that is—but more important, the relative loudness of different pitches alters. Sounds with long wavelengths, which correspond to low pitches, can refract or bend around
objects and keep traveling. Short wavelengths tend to bump into things
and get absorbed. The basic rule is that sound waves have difficulty ne-
gotiating an obstacle whose size is roughly equal to or greater than its
wavelength. So as a sound travels, it changes character.

Experiments with Carolina wrens have shown that these changes in
the relative strength of a song provide reliable cues to the distance the
signal has traveled. If you head into the woods, find a Carolina wren, set
up a loudspeaker, and play a tape of a Carolina wren song that was
recorded from a long way off, the real wren will sing back, but quickly
resume its business of foraging as soon as the tape stops. But if you play
a tape of a song that was recorded from close up, the real wren will immediately stop what it's doing and launch a ferocious attack on the
loudspeaker.

Further experiments confirmed that it was not the overall loudness
or softness of the song that cued in the listener as to how far off the
source was. The birds were clearly reacting to how distorted and degraded the signal was from the original. Moreover, a bird could only
make an accurate distance assessment if the song were one that the listener had in its own repertoire of songs. This last point is particularly
intriguing, for it is further evidence of a mental representation—one
that a bird uses not only for singing but for making ranging judgments.
It also clearly implies a sophisticated calculation is being made—not
just matching or categorizing or sequencing, not just judging whether
something is like another, but *how* alike it is.

Small birds such as wrens and tits need to spend almost all their waking hours feeding; they also need to defend their territory from intrusion. If they waste time and energy defending against birds that are too far away to constitute a territorial menace, they are going to drop dead from starvation. Likewise, if they fail to police their territory against actual invasions, it will soon no longer be their territory. Birds that can respond differently according to how degraded the incoming signal is are in the best position to optimize their response to the reality of the threat.

Much has been made in the pop-science literature about the discovery that a number of species (whales, most notably, but also some birds) have "dialects." The implication is that this represents something that surpasses even language ability in individual animals; it belongs to the realm of an animal "culture" or even, to New Agers and others in search of scraps of scientific-seeming proof of unscientific mysticism, to some higher group consciousness. Studies of Carolina wrens, however, suggest a perfectly sound and nonmystical evolutionary reason for the rise of dialects. When sender and receiver have aligned interests, vocal signals can converge on a stable, informative pattern; this was the case of the ground squirrel "hawk" and "badger" signals. In wrens, the sender sings to police its territory; the receiver takes advantage of the properties of the signal to decide whether to blithely keep foraging or to stop and rush out to defend its border. The sender has no particular interest in handing over such information most of the time. In fact, a wren that could make its signal's degradation difficult to read ought to have an edge. His neighbors would never be sure if he's near or far, and would always be needing to drop everything, run out and investigate, and with any luck drop dead from all of that needless rushing around.

Because birds can only accurately determine the distance of a song if they themselves possess that song in their repertoire, Carolina wrens have evolved an obvious strategy: use a lot of different songs, and keep changing them around in the hopes of always having a few that your neighbor doesn't. This is a typical evolutionary "arms race," an inherently unstable, noncooperative situation. What makes it so odd and fascinating is that the arms race is between members of a single species—indeed between the different roles of a single individual. As a receiver, every male Carolina wren tries to accurately measure its distance from another male. As a sender, it tries to thwart accurate distance measurements. Carolina wrens typically have thirty songs in their repertoire. Interestingly, there is one circumstance in which a male would want to send a signal that can be accurately ranged. This is in the situation that Morton refers to as an outright border war. Most of the time it pays to mess with your neighbor's mind by making your call seem close (or at a distance unknown and unknowable without time-consuming investigation). But when two neighbors directly confront one another over a boundary, the game has changed. In such a case, a bird wants to up the threat. And just as rutting elk engage in elaborate displays of vocal aggression or other rituals that express a threat and a willingness to fight—precisely as a way to avoid a lot of unnecessary energy expenditure and even bloodshed—Carolina wrens in this case *want* their proximity to be known loud and clear. Two wrens that engage in such a close-range cross-border dispute engage in what has been called "matched countersinging": one will switch its song to match the one the other is using. That is a way to ensure that your song *can* be accurately ranged.

So that explains the rise of ever more elaborate repertoires. But what of dialects? Dialects are stable but regional songs. By Morton's distance-measuring explanation, birds that exhibit dialects should have no need to disguise their distance from one another. They do not switch songs or play other tricks; everyone sings from the same sheet all the time. That explanation does indeed seem to match the evidence well. Birds that live in warm, stable areas with an abundant food supply, and where the same individuals tend to return to the same territories year after year, have little reason to challenge one another or engage in land grabs. These are the good-fences-make-good-neighbors types; all they need for keeping peace and order is clearly demarcated boundaries. The repertory singers in fact all do tend to live in colder, more northerly, and more food-scarce environments where mortality rates are high and competition for food is fierce. Red-winged blackbirds, most interestingly of all, show both tendencies depending on where they live.

In sunny California, blackbirds share a common dialect; in cold Wisconsin they each have as many as nine different songs.

The fact that dialects differ from region to region in the first place probably has a much simpler explanation. For one thing, different songs are acoustically better suited to different environments. For another, the regional differences may simply reflect random errors or "mutations"—just as American English came to differ from British English in many ways simply as a result of their isolation from one another. Many of the more complex regional variations found in birds with rich repertoires are also explainable through random changes. A simple mathematical model has shown that birds in different regions can develop their overlapping repertoires via a mechanism in which birds copy songs from one another with a low but constant random error rate. An apparently rich and seeming mysterious or even inexplicable "cultural" phenomenon is really just the product of chance. What the repertory-singing birds really need, after all, is variety, and they get it any way they can.

THE CASE OF THE SEMANTIC MONKEYS

Vervet monkeys do what ground squirrels do, plus 50 percent extra. In place of the two predator "alarms" of the squirrels, the monkeys have three. A vervet monkey that spots a leopard produces a loud, barking sound. If it spots an eagle it makes a chuckling bark. If it spots a snake it makes a high-pitched chatter. Just as with the ground squirrels, these different sounds elicit different antipredator behaviors. The leopard call sends all the monkeys running for the trees; the eagle call causes the monkeys to look up and run for the cover of bushes; and the snake call causes them to stand up on their hind legs and look down at the grass around them.

Experiments in which recordings of the calls were played back to monkeys in the wild when no predator was around were able to confirm that it was the call, and not the sight of a predator, that triggered the response in each case. Other experiments, in the lab, showed that primates' vocalizations are under relatively voluntary control, and are not just automatic responses triggered uncontrollably by a stimulus. Monkeys, for example, can be trained to make calls in response to certain conditions.

Perhaps because people are naturally more impressed by what monkeys do than by what squirrels do, the vervet monkey calls are frequently cited as a proven example of semantic communication by an animal under natural conditions. Dorothy Cheney and Robert Seyfarth, who conducted much of the field research, concluded that the alarm calls "functioned as representational, or semantic, signals." And because a monkey that hears these calls responds just the way a monkey that saw the predator himself would, "it is tempting to suppose that in the monkey's mind the call 'stands for' or 'conjures up images of'" the predator.

Cheney and Seyfarth and others have also drawn significance from the fact that young vervets make a lot of mistakes in their use of calls, but over time—apparently by watching how their mothers and other monkeys in the group respond—they get better. A young vervet may, for example, use an eagle alarm when it sees a leaf falling, or a snake call when it sees a hanging vine, or a leopard call when it sees a warthog. Young monkeys also tend to look to their mothers before responding to an alarm call. Thus they seem to learn through experience precisely what each alarm call means.

But as we saw in the case of the ground squirrels, there is another way of looking at these things. First of all, there is nothing remarkable, and certainly nothing necessarily linguistic or semantic, about an ability to form a learned association that causes an animal to react to an associated event just as if it had experienced the real thing itself. A dog runs to its food bowl when it hears the dog food bag being opened, a Carolina wren attacks the source of the sound of a nondegraded wren song. Even a worm can be taught which branch of a T-maze to choose to obtain a reward or avoid a punishment. Does the left branch of the maze therefore "conjure up an image" of an electric shock in the worm's mind?

If vervet monkeys are really able to "conjure up" the concept of a leopard or a snake, they show some remarkable limitations in their grasp of that concept. Vervet monkeys teach their young remarkably little about predators. They appear to understand nothing about the clues that predators leave in the environment, such as carcasses left by leopards in trees or the tracks left by snakes on the ground.

The argument that vervet monkey calls really are semantic rests strongly on Cheney and Seyfarth's account of how young monkeys learn to use the calls with increasing semantic precision. But Donald Owings suggests that a fairly simple observational bias on the part of researchers may have overestimated the number of "mistakes" that young vervets make in using the various calls. Even if they make very few mistakes, infants will stick out like a sore thumb to observers. Infants may be right an awful lot of time without there being any way to know it: When there is a real predator, the statistical odds are simply in favor of an adult being the first to call out, since all of them know to respond under those circumstances. In these cases, the infants never have a chance to demonstrate that they might have had the right answer had they been the first to call out. But when there is a mistake made, the infant is nearly always out there calling by himself. An infant may make one mistake for every ninety-nine correct calls, but it is the one mistake that will be noticed by an observer.

In other words, the calls may be more hard-wired than has been generally believed, less a matter of semantics than simple evolutionary self-interest, more a matter of calls that *do* than calls that *mean.* The appropriate response to each of the three kinds of predators requires a very different physical action. The effective response to a snake is "mobbing" of the kind ground squirrels employ against badgers; the effective response to an eagle is identical to the squirrel's every-man-for-himself-run-for-the-bushes anti-hawk response; the response to a leopard is much the same except that it's head-for-the-trees. The self-interest of the caller in each case is thus directly equivalent to the situation we saw in the ground squirrels—the caller benefits either by recruiting mobbers to help intimidate the intruder (snake, badger) or by setting off a display of multiple, distracting, rapidly moving targets that help take the attention away from himself.

Cheney and Seyfarth have more recently backed away from some of their earlier enthusiasm for their monkeys' cognitive abilities, and acknowledge now that monkeys appear to lack an ability to recognize the existence of mental states other than their own. Thus, their communication is not an "intent to modify the mental states of others." We may, however, be justified in viewing it as an intent to modify the *actions* of others. Evolution produces such behaviors all the time in social animals.

THE BARK PROBLEM

My dog, like most dogs, barks a great deal. She barks when she wants to come in, she barks when her dinner is late, she barks when the UPS truck comes down the driveway, she barks when she's accidentally left out in the field and can't get through the gate up to the house, she barks when the Border collie is herding the sheep and she's left outside the gate, she barks when she wants to play tug-of-war with an old rag.

If whines—over the course of eons of coevolution between sender and receiver—have come to "mean" fear or appeasement, and if growls likewise have come to mean aggression or a threat, what do barks "mean"? Acoustically, the bark shows an interesting characteristic: it is almost halfway between a growl and a whine. Barks rise and fall sharply in pitch, combining the rough qualities of a growl with the falling, tonal qualities of a whine. In terms of pitch they also fall halfway between the low-pitched growl and the high-pitched whine.

In the course of his survey of the sound spectrum of the animal kingdom (a task made easier by the fact that he had the National Zoological Park conveniently at his disposal), Eugene Morton found that many species bark, just as many whine and growl. Many birds' chirps are acoustically near-perfect barks—short, rising-and-falling sounds. If you tape-record a bird's chirp or chirt or cheep or chip and play it back slowed down, it sounds astonishingly like a dog bark.

The way these sounds are employed in the wild suggests that, evolutionarily speaking, these sounds have been selected for their task precisely because they are content-neutral according to Morton's motivational-structural rules. Birds and mammals will often use these sounds when they have spotted something of interest in their environment. The bark or chirp acts as a temporizing measure in this situation—a way to announce one's presence and wait to see what happens next without committing to a course of action. (Useful if anthropo-morphic analogy: A sentry who responds to every crash in the woods with "I surrender!" or "Charge!" is not going to last very long. "Halt, who goes there?" is a way to stall for additional information.) This is the way a dog uses its bark when, say, a car approaches. The bark may turn into a whine if the car turns out to be its owner, a growl if it's a stranger. In potentially hostile territorial encounters with a member of one's own species, a bark is in effect an announcement, "I'm here, wad-daya say to that?" It is a bad course of action to start growling before you know what you're growling at.

Precisely because the bark does not intrinsically mean anything the way a whine or a growl does, it can be pressed into service for a variety of tasks that are important to each species. Chickadees "chip" as they move to a new foraging site. Other chickadees in hearing range usually respond by moving along with the caller. A rampant semanticist would not hesitate to label this a "food call," or a "contact call." A more evo-lutionary-minded scientist would ask what's in it for the sender and re-ceiver to, respectively, make such a call and respond to such a call. The answer is pretty clear: A chickadee that is moving on to a new spot ben-efits from the mutual protection of keeping the rest of the flock moving along with him. A chickadee that hears such a call and moves benefits from sharing in the newly found foraging site. Thus a general-purpose, motivationally neutral call can, over the course of species-specific selec-tion, take on a more narrow function.

The vervet monkey "alarm" calls may well have been refined in this fashion. In the case of the ground squirrels, there was a specific acoustic property of each call—its localizability in space—that favored its se-lection for its evolved purpose. It is less clear whether that is the case with the vervet monkey calls. But one simple possibility is that the calls took the form they did originally because they in fact did accurately re-flect, according to the motivational-structural rules, the monkey's rela-tive state of alarm, with higher pitch reflecting greater fear. That does not mean that in their evolved forms they are mere "groans of pain" (or, perhaps, "squeals of fear"). But that original relationship could ex-plain why each has the acoustic property it now has.

Barks, though, are quite unburdened by any intrinsic acoustic "meaning." And that may explain why at least domestic dogs use them so liberally, and flexibly. As Mark Feinstein, a linguist at Hampshire College in Amherst, Massachusetts, has pointed out, precisely because barks mean nothing they can mean anything. In dogs, specific novel uses of barking thus do not represent an evolved use but rather a learned use. When a dog barks at the door, we usually let him in. When he barks at a strange car, we come out and look. When he barks at the closet where the dog food is kept, we feed him.

A bark is not a semantic word, nor is a vervet alarm call. But one might argue that each is a first step toward a proto-word. The dissociation between sounds and their intrinsic "meaning" that acoustics and evolution have conferred upon them is clearly a first prerequisite for human language. Words in all human languages have been freed of intrinsic acoustic meaning; it may well be no coincidence that words in all human languages are well-mixed blends of vowels (tonal sounds which share the basic acoustic properties of whines) and consonants (rough, nontonal sounds that share the basic acoustical properties of growls). Like barks, they intrinsically mean nothing, and so can be safely employed to mean anything.

A closely related question that people who own animals often wonder about is how much our dogs understand of our efforts to communicate with them. Most of the commands we teach dogs are readily explained as simple learned associations, though in fairness one should probably not call them so "simple." Dogs draw on many contextual cues to respond appropriately, and often what we say is the least of it. If you try the experiment of giving your dog a well-learned voice command over an intercom, you will find he usually does quite poorly. By the same token, if you say "fly clown" in the same tone of voice you use for "lie down," he will almost certainly not pick up on the difference. In fact, you can generally say anything you want and your dog will come (assuming you have taught him to come when called in the first place, that is). You can say "Here!" or "Come!" or "Tippy Tippy Tippy!" or "Afghanistan bananistan"; it makes no difference. What the dog is picking up on is your general tone of voice, your physical movements and gestures, and the general contingent circumstances. An interesting point here is that good trainers naturally use their tone of voice in teaching and giving commands—using intonation just the way that the motivational-structural rules imply is the most effective. When you want your dog to come, a high-pitched, tonal sound works best. When you want your dog to lie down, a rough, growl-like tone is most effective. The actual consonants and vowels are far less important. And since we tend to use the same precise commands in the same precise circumstances over and over, the circumstances offer an enormous source of cues. I can tell our Border collie "excuse me" when he is standing in the way of the door while I'm trying to open it to let him out, and he will shoot out of the way like a rocket. But if I say "excuse me" in an entirely different situation, I generally get no more than a blank stare.

Likewise, dogs "know" their name only in the sense that they have associated that sound (or any rough equivalent, per above) simply with the equivalent of the command "here" (we say their name, they come to us, we reward them). Or perhaps more often, we use their names as a sound that simply means in effect "pay attention" to what follows. (In this sense a "name" almost perfectly takes on the "bark" function—a sound that serves to mark interest and elicit attention. Thus we say their name, *something* important follows—a reward, another command.) Gary Larson summed this up perfectly in his "Far Side" cartoon showing a person yelling at a dog, with the following explanatory caption:

WHAT WE SAY TO DOGS: "Okay, Ginger! I've had it! You stay out of the garbage! Understand, Ginger? Stay out of the garbage, or else."

WHAT THEY HEAR: "Blah blah GINGER blah blah blah blah blah blah blah blah GINGER blah blah blah blah blah."

TALKING APES

This is all no doubt disappointing to pet owners who are convinced that they really are talking back and forth to their dogs or cats. But maybe it is just as well we cannot really talk to animals: Consider Nim, a chimpanzee who was taught a simple form of sign language through years of intensive training (and lots of bananas and candy as rewards). Here is what Nim had to say on a typical day: "Nim eat, Nim eat, Drink eat me Nim, Nim gum me gum, You me banana me banana you."

Still, even a very boring talking chimpanzee commands our attention; it surely falls in the class of dogs that ride bicycles, where the impressive thing is that they can do it at all. It would be uncharitable for us to carp that the dog has not yet won the Tour de France.

That assumes, of course, that the dog really is riding the bicycle. If it turns out that its trainers had tied the dog's paws to the pedals and added training wheels and a motor—and then argued that the dog was nonetheless demonstrating the essential elements of bicycle riding since its legs moved up and down and the bicycle moved—we might take a different view of the matter. Nim's story has become a classic in the history of animal language projects. It marked a turning point in how scientists look upon such endeavors and what they mean. Nim's trainer, Herbert Terrace, was following on several other ape language projects in the late 1960s and 1970s. A much earlier attempt to teach a chimpanzee to literally speak had proved an utter failure. In the 1940s, a baby chimpanzee named Vicki was raised in a human household by two psychologists who tried to give the chimp all of the experiences a human child would have. Human babies acquire speech spontaneously and the hope was that Vicki would, too. Human babies do not, however, undergo arduous sessions in which their mouths are manually forced into various shapes by their parents to teach them how to say sounds. Even with such training, Vicki never was able to produce more than four words that anyone, even with great goodwill and indulgence, could recognize as an English utterance. Besides possibly saying *mama*, *papa*, *cup*, and *up* (sometimes Vicki would use her own hands to shape her mouth to produce these words), Vicki could carry out simple commands to fetch objects.

Ape-language researchers who followed this failure recognized that chimpanzees simply lack the vocal apparatus to form the sounds that make up human words, and so decided to teach what they hoped would be a more natural form of communication, American Sign Language.

These projects generated much attention, not a little of it self-generated by the research teams themselves who made appearances on the *Tonight Show* and a National Geographic Society television special. After three years of training, Washoe, a chimpanzee, was reported to use sixty-eight signs, and to form a number of two-word and even three-word combinations (*you/me/go out*) in ways that resembled the way a human child starts to form sentences. "It was rather as if a seismometer left on the moon started to tap out S-O-S," the linguist Roger Brown excitedly commented. "Language is no longer the exclusive domain of man," declared the trainer of Koko, a gorilla who performed similar feats.

Terrace sought to repeat these experiments himself. Nim (short for Nim Chimpsky, a humorous jab at linguist Noam Chomsky's contention that language is a uniquely human phenomenon) was raised in a highly social setting with a number of human volunteers who encouraged him to help with household activities such as washing the dishes, and was taught a series of ASL signs. Terrace also extensively videotaped Nim's signing sessions, and had completed analyzing his data and writing the first ten chapters of a book reporting Nim's success in acquiring language when, in the course of "some routine checks" of the tapes his work was "interrupted by a dramatic and unexpected discovery." Slow-motion playbacks of the tapes, Terrace found,

> revealed that most of Nim's signing was prompted by his teachers and that, worse still, his signing was largely imitative of what his teachers had just signed to him. In contrast to the largely spontaneous and creative use of language by children (hearing and deaf alike), Nim's signing turned out to be a means toward the end of obtaining some reward from his teacher. The videotapes also showed that signing was a means of last resort. Typically, Nim tried to obtain his reward by direct physical means. It was only when those efforts failed that Nim responded to his teacher's urgings to sign. . . . The very tape I planned to use to document Nim's ability to sign provided decisive evidence that I had vastly overestimated his linguistic competence.

For example, if Nim tried to reach for his pet cat, his teacher would try to get Nim to "converse" about the cat by signing a question such as *Where cat?* When Nim responded he was using a sign that his teacher had just used. Most of Nim's responses, the tapes revealed, would consist of the sign the teacher had just used plus some "universally relevant signs" such as *me, more, Nim, hug* to form a "sentence." Looked at in isolation, Nim seemed to be forming appropriate and novel sequences of signs that took the form of grammatical sentences, combining an agent and an object *(me gum)*, an agent and an action *(me eat)*, an action and an object *(tickle me)*, and so on. Indeed, Terrace noted that the "unspontaneous and imitative" nature of Nim's signing was not apparent at the time or even during the hundreds of hours spent preparing transcripts of the tape. It was only through painstaking frame-by-frame analysis of the tapes that the true story became apparent.

Similar analysis by Terrace's team of publicly available tapes of Washoe and Koko showed a similar phenomenon—the apes were largely imitating signs that their trainers had just used. Worse was the degree of wishful thinking that became apparent in some of these projects. Koko's many mistakes were explained away by her trainer as metaphors or mischievous lies; the Washoe research team, according to a subsequent revealing statement by the one native ASL signer on the team, was extremely generous in its interpretation of the chimpanzee's signs. "Every time the chimp made a sign, we were supposed to write it down in the log," he recalled. "They were always complaining because my log didn't show enough signs. . . . I just wasn't seeing any signs. The hearing people were logging every movement the chimp made as a sign. Every time the chimp put his finger in his mouth, they'd say 'Oh, he's making the sign for *drink*,' and they'd give him some milk." When Washoe would reach for something, the hearing members of the team would say, "'Oh, amazing, look at that, it's exactly like the ASL sign for *give!*' It wasn't."

Terrace is still accused by ape language enthusiasts of being a closet Skinnerian deliberately out to debunk the whole field. If so, he must have been a peculiarly conspiratorial one to have pretended to be an enthusiast himself for years in order to stage a last-minute reversal. A simpler explanation is that he was willing to admit he was wrong. "I know what it's like," he told the *New York Times* in 1995. "I was once stung by the same bug. I really wanted to communicate with a chimpanzee and find out what the world looks like from a chimpanzee's point of view."

The role of spoilsport is not a particularly enjoyable or popular one, and enthusiasm for the hope and belief that apes really can be taught communicative, humanlike language remains hugely popular among the public at large, even as it has acquired an overall bad odor among most cognitive scientists. A popular book for children about Project Nim was written by one of Terrace's volunteer teachers and doesn't deviate a whit from the happy and acceptable story line that an ape really was taught to communicate. "With enough sign language a chimpanzee might someday tell us much—about his past and his future, his feelings and his dreams. Who knows what a chimpanzee might say if he learned the words?" the book gushingly concludes. Even Terrace himself could not apparently bring himself to dash this tone of childlike wonder; in the introduction that he contributed to the book he ducked into what might be seen as a rather lawyerly evasion by saying, "To all of us who worked and communicated with Nim it seemed obvious that he was creating his own sentences in much the same way that a human baby would." *Seemed obvious*, yes; that was Terrace's point when he announced what the painstaking reanalysis of the videotapes actually revealed. The most Terrace could bring himself to say in the children's book was, "But I still didn't have the conclusive evidence to prove scientifically that Nim's sequences of signs were actual sentences."

CLEVER KANZI

There is more than a touch of Clever Hans lurking behind these experiments: The apes' trainers were unconsciously providing the cues that triggered the apes' responses. But there was something else, too, namely a failure to distinguish between the act of *using* a symbol and the ability to *understand* its symbolic, or referential, nature. The most recent ape star is Kanzi, a pygmy chimpanzee or bonobo trained by researcher Sue Savage-Rumbaugh. (Bonobos are considered to be either a subspecies of "regular" chimpanzees, or a separate though closely related species.) Instead of sign language, Kanzi uses "lexigrams," which are abstract geometric symbols that are presented either on plastic tiles or on a computer keyboard. In her earlier chimpanzee language experiments, Savage-Rumbaugh and her coworkers taught three chimpanzees, Lana, Sherman, and Austin, to use lexigrams to form "sentences" in a prescribed order to get rewards. Pressing the keys that stand for *please/machine/give/banana*, for example, would yield a banana.

In one series of experiments, Sherman and Austin were shown a selection of five to seven different foods on a table, then were required to go to another room, choose one lexigram, then return to the table and take one food and show it to the experimenter. The chimpanzee would be allowed to eat his selection only if it matched his chosen lexigram. Together, Sherman and Austin made the correct choice of objects 80 percent of the time.

Kanzi's much more impressive feat was to acquire the use of such symbols without any training at all. For most of his first thirty months of life, Kanzi was kept with his foster mother Matata in a room where she was (unsuccessfully) being taught the use of lexigrams. When Matata was returned to the breeding colony and Kanzi left alone in the room, Kanzi immediately began using the lexigrams to make requests of his human companions for food, toys, and activities like tickling and chasing.

All of these feats testify to chimpanzees' impressive cognitive abilities. Sherman, Austin, and Lana all learned to manipulate multiple symbols. Kanzi's learning by observation as Matata pushed lexigrams and received various rewards is one of the few thoroughly documented instances of observational learning in animals. It remains far from clear, however, that any of this use of symbols equates to an understanding of symbols. Terrace was making a deliberate point about this when he trained pigeons to peck out four-item sequences of colored lights to obtain rewards. How, he asked, was that any different from Lana's ability to choose a sequence of lexigrams that her human experimenters *chose* to interpret as standing for "please machine give banana"? There is indeed a semantic trick in even calling one of those lexigrams "please." Why "please"? All that key does is initialize the computer. Does the ape understand the concept of "please"? No better than the pigeon would understand that pecking the blue key means "please," Terrace argued. Lana had to get the four lexigrams in proper grammatical order. Does that mean she understands the rules of grammar? Again, no more than the pigeon does. The very fact that human researchers chose to label these lexigrams as words with precise semantic or grammatical functions biases the entire interpretation of what the animals have done.

Dolphin trainers have made much of the claim that dolphins have an understanding of grammar, evidenced by the fact that they can learn to distinguish between commands such as "take the Frisbee to the surfboard" and "take the surfboard to the Frisbee" given in sign language or sound symbols. "If you accept that semantics and syntax are core attributes of human language, then we have shown that the dolphins also account for these two features within the limits of this language," says Louis Hermann, who directs a dolphin language project in Hawaii.

But is this grammar, or is it again just rote learning? We have already seen many examples of how apes, monkeys, and even pigeons to some extent can learn sequences and transitive properties of lists. And in interpreting even complex syntactical strings of lexigrams or hand signals or sound signals, there are plenty of shortcuts an animal can take that involve no understanding of the underlying rules of grammar. A chimpanzee named Sarah was taught to use lexigrams and could reportedly learn to distinguish these two sentences:

If / Sarah / take / apple / then / Mary / give / Sarah / chocolate, and
If / Sarah / take / banana / then / Mary / no / give / Sarah / chocolate

But Sarah really didn't have to notice very much about these two strings of lexigrams to catch on about what do to—really nothing more than "apple" in the first sentence and "banana no" in the second. Since every task she is given results in a reward, the "If" and the "then Mary give Sarah chocolate" parts of each string can be totally ignored. Sarah doesn't even need to notice the "take" lexigram because the smallest amount of trial and error will reveal that taking (a not unnatural act for a chimpanzee) is what the game is about. A dog could certainly be trained to learn an equivalent task ("get your ball" versus "get your Frisbee" versus "Frisbee no"). Likewise, the dolphin that learns by experience that the first named object is the one to fetch is not exactly displaying any deep grammatical knowledge. It is displaying an ability to form a learning set and transfer it to novel examples—but again cognition is not language.

It is always a revealing exercise to think about how we would interpret a claim for animal language ability if we had done exactly the same experiment without labeling as "words" the symbols the animal has been taught to use. No one would even think of a categorization or matching test as a measure of "linguistic" ability. Pigeons were taught to peck one of four keys according to whether the picture they were shown was a cat, a car, a flower, or a chair. An orangutan at the National Zoo in Washington, D.C., has been taught in the course of the "Orangutan Language Project" to pick one of six keys according to whether the object (or a picture) it is shown is a banana, grapes, an apple, monkey chow, a cup, or a bag. Not coincidentally, the orangutan gets to eat a banana, or grapes, or an apple, or monkey chow when it gets the right answer for the food items. Also, perhaps not coincidentally, he seems much less interested when it's a bag or a cup, and takes significantly longer to answer those questions. The only difference between what the pigeons have done and what the orangutan has done is that one is called a categorization test and the other a "language project."

Although Savage-Rumbaugh has been scrupulous in eliminating one of the major faults of Project Nim—the provision of unconscious cues to the animals, which they simply imitate in their responses—the fact remains that the lexigram-using chimpanzees are still nearly always producing their "sentences" only when they want something. Savage-Rumbaugh reports that 96 percent of Kanzi's utterances take the form of demands for food, toys, tickles, or other activities. As we shall see, that fact alone implies something fundamentally different about what Kanzi is understanding and what a human child is.

THE NAME

Using a symbol and understanding a symbol are not the same thing. I can write a very simple computer program that could reproduce all of the decisions that Kanzi or Sarah makes in creating or reacting to lexigrams. The Talking Moose can manipulate symbols, and even turn them into vocal signals, without understanding a thing about them. A pigeon can peck a sequence of colored lights without having a linguistic concept that each light stands for something.

When an ape uses a symbol it has been taught, it invariably is using that symbol as a means to a very tangible end—a means to request an object or a pleasurable activity. But human children from the earliest ages use the symbols of language in a fundamentally different way. Most striking is the fact that right from the start of their efforts to communicate, children use words as *names*, not just as requests. Toddlers as they learn words immediately begin to use them not only as "protoimperatives" but also as "protodeclaratives," for no other purpose than to demonstrate their understanding of the name and to call a parent's or other person's attention to an object of interest. Children will say "red" or "plane" or "cat" in circumstances where they show no interest at all in obtaining the object referred to. Uttering the name is a goal in itself. This behavior appears spontaneously at around nine to thirteen months. Children go on to quickly acquire new names of objects with minimal repetitions or exposures to exemplars of the noun.

The huge difference in the sheer number of symbols that humans use in human language versus the number that apes can acquire in training—not to mention the automatic ease with which humans acquire words versus the extensive training with potent rewards required to teach apes—suggests that the difference is not one of just degree (say memory capacity) but of fundamental underlying processes. One of the fascinating things that arise from studies of how children acquire language is that very closely tied to an ability to use words as names— that is, as symbols that refer to a concept, and not just as means to demand a desired end—is a "theory of mind" about others. It is the intention to share attention to an object with another that lays the groundwork for using words as names. Even before an infant begins to use words, it will avidly engage in interactions with its mother that involve directing attention jointly to an object. As early as the fourth month, a child will follow the gaze of its mother to look where she is looking. Later, usually between eight and twelve months, children learn to point to objects themselves to call their parents' attention to them: They have learned that it is possible to communicate *about* things in the world *to* other people. The psychologist Simon Baron-Cohen points out that by this very act toddlers are seeking to influence another person's attention as an end in itself. They are seeking "to affect the listener's *mind*." The urge to communicate and the discovery that objects have names arrive simultaneously within this remarkably short, and remarkably early, window in human development.

Many games that parents and children instinctively play with one another have the effect of reinforcing this idea that communication and naming is fundamentally about telling *someone else* about something of interest in the world. A mother will pick up an object and show it to her baby, or point to something and say its name. Peekaboo games also work to reinforce this concept of joint attention to an object for its own sake. Infant apes, by contrast, invariably approach a new object from a purely acquisitive point of view, taking it in their hands and investigating it with their mouths and fingers. But in a peekaboo game, the object's appearance and disappearance serves to make the object itself, and the parent's and child's joint attention to it, the focus. Terrace notes that infants learn through these games not only to join a parent's attention to an object, but also that its own response—such as pointing or babbling or laughing or eventually saying its name—is recognized by the parent as a sign that it has noticed the object. "In short, the infant learns that her or his response to an object has much in common with the parent's response to the same object," Terrace says. Blind children obviously lack the ability to engage in any visual joint-attention games, but nevertheless develop with their parents a mutual attention toward features of the environment that they can sense. And it is interesting that even blind children quickly and naturally learn to express this idea of calling someone else's attention to an object of mutual interest with the word "look."

Apes, by contrast, rarely manipulate objects with the aim of engaging their infants' attention, nor do infant apes attend to the manipulation of objects by their parents. Nor do apes under natural conditions, except very rarely, point at objects.

One experiment did seem to show that a chimpanzee could name an object even when there was no immediate reward in sight. In a variation on the name-the-food-you're-going-to-pick test, Sherman and Austin were shown a table containing five randomly picked inedible items—tools, photographs of tools, and photographs of food. Again they had to go into the next room and pick the lexigram that corresponded to one of the objects, then return and pick out the object they had indicated and hand it to the experimenter. They got 90 percent right, even though all they got for a right answer was praise. This could be seen as a unique example of a nonhuman animal naming and directing attention to an object without an instrumental end in sight. But of course the chimpanzees had already done an almost identical experiment where they *did* get an immediate food reward. (When I taught my dog to bark on command I gave him a dog biscuit as a reward, but now he will do it almost every time without a reward.) Indeed, the chimpanzees' behavior in this experiment with inedible objects included a very telling wrinkle: After indicating symbolically which object they were going to present to their teacher, they "would again point, now deliberately and with a more expressive gesture, at the object they just named . . . [they] clearly expected some sort of reward (either food, tickling, or praise) for having named and pointed to an object."

Much of the futile definitional debate over whether language-trained animals are displaying language has focused on technical definitions over what constitutes semantics and syntax. Of course, one overarching point that often gets lost is that none of these animals displays any such abilities at all in the wild. Even assuming that language training experiments reveal something about inherent linguistic abilities in animals, this definitional debate really misses the point. What matters is not what an animal is doing in a few trivial examples; what matters is how it does it. It is the fundamental difference between using symbols and understanding symbols that constitutes the discontinuity between animals and humans, and which leads to the manifest and huge gap between the rote demands of language-trained apes and the conceptual flights of humans. Savage-Rumbaugh has argued that in terms of quantitative acquisition of "words" and combination of those words into sentences, Kanzi equals the abilities of a two-and-a-half-year-old human. But the qualitative difference in the way Kanzi employs words from the way even a year-old child does is vast, and is what allows a child to keep on expanding its grasp and use of language. "If a child did exactly what the best chimpanzee did," says Terrace, "the child would be thought of as disturbed."

7

ANIMAL CONSCIOUSNESS

◀◀ DRESSING A CHIMPANZEE IN DIAPERS, SENDING HIM to a good school, and bribing him with cartloads of bananas and M&Ms still doesn't make him into a chimpanzee who can use language the way we do. As we have seen in chapter 6, the circumstantial evidence from studies of language development suggests that this difference is a fundamental one, having to do with a unique human awareness of self and other. Animals use communication in a fashion that appears to seek influence over the behavior of others, but not the thoughts or knowledge of others. Humans, by contrast, from infancy show an understanding that other humans have minds that work roughly the same way their own do, and whose knowledge can be altered by words and actions.

Many studies have sought to probe more directly whether other animals possess a "theory of mind," which we can define as a grasp of the notion that others have thoughts and beliefs that may differ from one's own. We have already encountered many examples of the extraordinary things animals can accomplish without recourse to conscious thoughts, beliefs, and intentions as we know them. Evolution, learning, the very wiring of animals' brains and sense organs, adapt them to the cognitive demands of their physical and social environments in ways that at times put us to shame, with our reliance on consciousness and language and reason to see us through. We can study books about cryptic moths and make long lists of their distinguishing characteristics and actively try to commit them to memory and in time can do as well as every blue jay in the world can. We might with enough mnemonic devices or other memory tricks find caches the way Clark's nutcrackers can. We can never follow a trail the way a dog can, or find a rotting carcass the way a turkey vulture can, or navigate our way home the way a pigeon can, or locate a tiny rapidly moving target the way a bat can, or calculate the distance of a Carolina wren from the tonal shifts in its song the way a Carolina wren can. Nor do we have any particular need to do these things (nor would we even want to be able to do some of them). So one might well argue that animals do not need the special human cognitive abilities that we possess, for they have gotten along quite well without them.

All of which is to say that there is a certain flavor of anthropocentric bias in the very hunt for self-awareness in other animals, a hint that conscious self-awareness is the best thing evolution has yet to produce—and we want to know how animals stack up against this standard of ultimate perfection. Yet it is no insult to animals that they might do what they do without self-awareness as we understand it; nor is it a particular compliment to animals to see how closely they share our peculiar cognitive abilities.

THINKING ABOUT THINKING

Still, there is no denying our natural curiosity about what's going on in there. An awareness of our own thoughts, and a curiosity about the thoughts of others, are as basic as human characteristics get. Having a concept of one's self as a conscious moral entity, and of others as having thoughts and an internal mental existence we seek to touch, are to us the essence of love, and life, and existence. When it comes to our pet animals, we are naturally uneasy with the suggestion that our dog is, in effect, faking it. Its behavior so clearly seems to express love, and concern, and pleasure in our company; if that behavior is not rooted in an awareness of itself and of us as conscious agents, that gift of love comes dangerously in peril of losing its meaning.

There are deep philosophical waters here. What self-awareness means is not easy to define. It is perilously close to the question of what consciousness is, which all prudent writers on this subject will leave to the philosophers, who have been arguing the point for two thousand years and are still going strong.

In one sense self-awareness may simply mean an ability to maintain a mental representation of one's own physical dimensions and internal physiological states. Even that raises philosophical tangles; is an animal that is hungry and seeks food "aware" that it is hungry? Can it somehow describe its state as one of hunger, separate from its actions that respond to that state?

Self-awareness may more properly describe an ability to grasp a concept of one's self and one's thoughts and to reflect upon these— much as humans grasp a concept of word-symbols as names, and of numerical labels as numbers, in a way that animals apparently do not, even if they have an ability to *use* these symbols and labels. Determining if an animal has a concept of words or numbers or categories has proved hard enough; determining if it has a concept of itself and of others as mental beings pushes the very limits of what may be philosophically knowable.

But we may still be able to probe the degree to which animals, by their behavior, show some ability to mentally represent their own state of mind, and to make decisions based on those representations. This question is at least a bit more tractable; it is a question of whether an animal's own mental states are *accessible* to itself, and it possibly is a separate question from the one of consciousness. In other words: How much does an animal seem to know about its own cognitive processes? Note that such self-awareness may still not necessarily imply consciousness as we experience it; unconscious thinking may include unconscious thinking about one's own unconscious thinking.

The ability to have thoughts about thoughts—"secondary representations" or "meta-representations"—does however represent a major threshold. It allows an organism to go beyond perceived reality, into the realm of the hypothetical. Representations of representations are what make insight possible; they are what allow an organism to try out possible solutions to a problem in its mind; they are what permit comparisons of past mental states to present, novel situations. Such a self-awareness may also be an essential prerequisite to other-awareness.

Being able to treat thoughts as independent entities—entities that can be cognitively stored and manipulated, separately from the things those thoughts represent—is the basis for recognizing that others can have thoughts and beliefs about the world that differ from our own. The psychologist Nicholas Humphrey suggested a number of years ago that the evolution of self-awareness in humans may have been an adaptive strategy precisely because it provided the material for speculating about the thoughts (and thus the motives) of others. In everyday life, we use this ability all the time in our dealings with fellow human beings. If we can speculate about what might be going on in someone else's mind, we can better predict their actions. If we can represent and access our own knowledge, we can understand the effect and consequences of withholding information from others. We can grasp the possibility that someone else may be deliberately deceiving us. We can devise ever more devious stratagems for manipulating the behavior of others by manipulating their thoughts and beliefs. The history of the kings of England and the goings-on of any large bureaucratic office provide copious illustrations of the utility of such Machiavellian intelligence in the affairs of men. Whenever we can successfully answer for ourselves the question, "What would I do in his position?" we are better able to anticipate and counter the stratagems of our rivals.

(The way people make such assessments of others' thoughts and motives also perfectly illustrates the fact that we are fundamentally drawing upon our own introspections as the model for the hypotheses we develop about the thoughts and intentions of others. We are all familiar with the conspiratorial-minded person who sees a conspiracy behind what everyone says, the manipulative person who sees manipulation as a universal explanation for human behavior, and for that matter the economist who assumes that everyone is exclusively motivated by money and self-interest.)

Many animals observe and act upon other animals' behavior, and likewise act in ways themselves that seek to influence others' behavior. But evidence that animals seek to penetrate the thoughts and beliefs of other animals, and to seek to influence those thoughts and beliefs, are harder to come by. As we will see, the evidence that monkeys have such a "theory of mind" is generally negative; the evidence for apes is mixed.

But we are getting ahead of the story, for the first question we need to examine is how well animals are aware of themselves, and of their own thoughts and knowledge.

WHAT ANIMALS KNOW ABOUT THEMSELVES

Animals don't generally bump into things. I have noticed that my horse is particularly adept at calculating clearances through narrow spaces; he can maneuver between two trees or around obstacles in a way that clearly reflects a knowledge of his body dimensions. A dog doesn't have any trouble knowing where to scratch itself to get a flea. Physical self-awareness is pretty much a universal trait among animals. Much of this may be a matter of simple learned associations between visual or tactile sensations and the ensuing consequences. Puppies do seem to bump into things a lot, for instance, and it is not difficult to imagine that they learn not to simply through experience—just as we learn to catch a baseball through visual and tactile feedback until the action becomes automatic. This sort of "self-aware" behavior probably tells us little about whether animals have representations about their own representations.

More revealing are experiments in which rats were trained to "report" on their own behaviors. The rats learned to press one of four different levers depending on what they were doing at the time a buzzer sounded—walking, standing, rearing up on their hind legs, or face-washing. (A correct response generated a small food reward.) Rats naturally perform all of these behaviors, and the buzzer would only sound while they were already engaged in these behaviors: so the experiment was not teaching the rats to *do* things, but rather to report on what they were already in the midst of doing.

Another experimental tactic for assessing an animal's awareness of its own mental state is to see whether animals know when they don't know the answer to a problem. The psychologist J. David Smith and his colleagues set up the following situation. Rhesus monkeys that had been previously trained to use a joystick to move a cursor around a computer screen were presented with a screen showing a box and an "S" symbol. The brightness of the box could be varied by changing the number of pixels within the box that were lit up. If the box was at its maximum brightness, the monkey's correct response was to move the cursor to the box. If the box was at anything less than that brightest intensity, the monkeys had to move the cursor to the S. Choosing the right answer earned an immediate reward; choosing the wrong answer triggered a delay before the next trial began.

The monkeys were started off with very easy problems, in which the dim, or "spare," patterns that they were presented were very much less bright than the maximum-brightness pattern. After about four hours' practice the monkeys had mastered the task. Then the experimenters gradually made the problem tougher—the spare patterns were made increasingly bright, so it became increasingly difficult to tell which of the two possible answers was correct. At this point the experimenters gave the monkeys a third option to choose. In addition to the box and the S, a star pattern was added to the screen. If the monkeys chose the star pattern, they would immediately be given a "guaranteed-win" problem—a screen with only a box or an S (much as Groucho Marx on his quiz show "You Bet Your Life" used to always offer losing contestants a consolation question such as, "When was the War of 1812?" or "Who's buried in Grant's Tomb?"). But repeated use of the star option would introduce increasing delays before the next question, so it would pay for the monkeys to select this bailout only when their ability to pick the right answer began to approach the 50–50 chance level.

That is precisely what the monkeys did. The peak use of the star option coincided almost precisely with the point at which the monkeys were equally likely to react to the spare pattern by choosing the box or choosing the S when they did try to make a choice between the two.

The same experiment was tried on human subjects, with almost identical results. The humans got points (redeemable in cash) for getting right answers or for choosing the bailout option; just as with the monkeys, however, choosing the bailout would introduce increasing delays before the next problem. Graphs of the choices the subjects actually made—box, S, or star—plotted against the difficulty of the problem were virtually indistinguishable for people and monkeys.

A similar experiment in which people and a dolphin had to tell whether a tone was higher or lower than a target tone presented at the beginning of the test yielded the same findings. The dolphin had three paddles to press—higher, lower, and uncertain—and it increasingly chose the bailout option as the tone being presented got closer to the target. The dolphin showed clear signs of indecision in the more difficult trials as well: it would approach the paddles more slowly and sweep its head from side to side before making a selection.

Subsequent interviews confirmed that the people who took the test were monitoring and reacting to their own subjective state. When they chose the box or the S, they explained their choice in terms of the objective, physical features of the box pattern. But when they chose the bailout option, the researchers noted, the human subjects always explained it in terms of their subjective mental state—"I just wasn't sure." (They didn't say "the density of the pattern was too close to the target" or "I noticed I was getting a lot of wrong answers.") The monkeys' and the dolphin's bailout responses would likewise seem to reflect "metacognitive reactions to subjective uncertainty," the researchers concluded—"that is, monkeys bail out when they know they do not know the solution of a trial."

As ingenious as these experiments are and as impressive the results, concluding that they establish self-awareness in monkeys is, as always, debatable. For example, it is perfectly plausible that the monkeys bail out simply when they start getting a lot of wrong answers; they could be drawing conclusions not from their subjective state but from observing the objective results. At a certain point in the course of the test, in other words, the star response simply works better. The ability of monkeys to learn new discriminations based on a very few reinforced trials could explain even a very rapid shift in strategy in response to a change in the reward pattern.

WHO'S THAT MONKEY IN THE MIRROR?

The most often cited evidence that some animals possess self-awareness and a concept of self—even an awareness of their own existence and (according to one published claim) an ability to contemplate their own mortality—are experiments in which apes appear to recognize themselves in mirrors. This phenomenon was first reported in detail in 1970 by researcher Gordon Gallup. He found that when chimpanzees were first exposed to a mirror they would react to their image as if it were another chimpanzee, with vocalizations and threatening gestures. Monkeys, dogs, and many other animals do the same; many will at first try to look or reach behind the mirror. But in time the mirror image loses interest for them, and they come to simply ignore it.

Chimpanzees, by contrast, appeared to progress from viewing the image as a fellow chimp to viewing it as themselves. They would stand before the mirror and groom visually inaccessible parts of their bodies, pick food from between their teeth, pick their noses, make faces and stick out their tongues, or just seem to use the mirror to visually explore parts of their bodies that they normally could not see—the insides of their mouths and the anal-genital areas.

Gallup then tried to test mirror-recognition ability in chimpanzees more systematically with what has come to be called the "mark test." Chimpanzees were first exposed to a mirror for ten days. On the eleventh day they were anesthetized and marked on the ridge of one eyebrow and on the top half of the opposite ear with a bright red dye. The chimpanzees were then watched to see what they would do when they came to. First they were placed in their cage without the mirror. On average they touched the marked areas only once during a half-hour period. Then the mirror was introduced, and the chimpanzees on average touched the marked spots seven times in a half hour. Some of the chimpanzees were reported to touch the marked spots while looking at their image in the mirror, then sniff or examine their fingers.

Gallup has made increasingly strong claims for the significance of these findings, arguing that self-recognition requires a mental concept of self. His research has spawned a mini-industry in mirror-recognition studies in the decades since. The mark test has been repeated on many different species, and the results, at least at first, seemed consistent with the idea that self-recognition constitutes a higher cognitive ability shared uniquely by man and the great apes. Orangutans, another member of the great ape family, were found to pass the test; gibbons and the many species of monkeys that were tested all failed.

Further testing with chimpanzees, however, yielded less consistent results than originally reported. A study with eleven chimpanzees found only one who touched the mark during the test. Three of the chimpanzees who did not touch the marks did, however, show the sort of "self-directed" behaviors Gallup first reported, such as using the mirror to examine themselves.

One very basic problem in interpreting the results of all of these experiments is that chimpanzees engage in these very same self-directed behaviors routinely, whether there is a mirror around or not. Chimpanzees very frequently touch their heads and faces; indeed, in the experiment with the eleven chimpanzees, one animal confounded the test by his tendency to rub his brow vigorously while coming out of the anesthesia—thereby rubbing the mark off even before he had the chance to view it in a mirror.

Another basic problem is that some of the very behaviors categorized as self-directed are also social responses that chimpanzees exhibit in the presence of a fellow chimpanzee. Self-grooming in many primates is a social behavior. Stumptailed macaques placed in a cage with a mirror along one wall show a marked increase in self-grooming. But so do stumptailed macaques placed in a cage next to another member of their species behind a transparent partition. So it is not always easy to tell from its behavior whether an animal is reacting in a "self-directed" manner to a mirror image as an image of itself, or "socially" to the image as an image of another animal.

Some (though not all) of the results reported as evidence of mirror recognition may even have been nothing more than an artifact of the way the experiments were carried out. One fact that Gallup originally reported was that upon coming out of anesthesia, the chimpanzees spent four times as much time viewing themselves in the mirror as they had on the day before the experiment. But this may be rather simply explained by the fact that the image in the mirror, which had been more or less the same for the previous ten days (and presumably rather boring) suddenly became far more interesting—it suddenly had a large red blob on its head.

In Gallup's original study, two chimpanzees that had not been exposed to mirrors at all were anesthetized and marked and confronted with a mirror for the first time, and did not show the same attempt to touch their marks that the experimental group showed. That seemed to be convincing evidence that the experimental group was indeed using the mirror to guide their manual explorations of the marks, because the only difference between the experimental and control groups was the previous experience with mirrors and the chance to learn that they constituted a self-image. But as the psychologist C. M. Heyes has pointed out, the control group may have been too busy reacting to its mirror image as a threatening fellow chimpanzee to engage in the normal grooming activity that (by pure chance) could have produced the instances of mark-touching observed in the experimental group. All chimpanzees react to mirrors on first exposure this way, and the control group's first exposure (right after the mark was applied) was no different from the experimental group's first exposure (ten days earlier). The only difference was that, with ten days' experience, the experimental group had already had time to stop reacting aggressively to their mirror images and get back to a more normal repertoire of behaviors.

Meanwhile, more-careful tests with monkeys have suggested that what the "higher" chimpanzees and orangutans are doing may be nothing cognitively very special at all. The cotton-top tamarin is a New World monkey species that has a distinctive white mark on its head. Earlier tests in which these monkeys were given the standard mark treatment elicited little or no interest in the monkeys' mirror image. But in a more recent study the experimenters tried dyeing the entire white tuft a shocking color—pink, purple, green, or blue. The monkeys treated in this fashion stared into the mirror for considerable periods of time and touched their hair while looking in the mirror. A control group was never observed to do the same. And the treated monkeys reacted to their own shocking mirror image very differently from the way they would react to seeing a fellow monkey with a dye job. Even though staring is usually a hostile threat among monkeys, and even though monkeys in other tests continued to show hostile reactions to their mirror image for months and even years, the dyed tamarins looked peacefully at their image for unusually long periods and did not vocalize or act aggressively toward it.

Marc Hauser, who conducted the experiment, suggests that if a monkey—a species shown by many other measures to lack a "theory of mind"—can do it, then equating self-awareness to self-recognition may be a red herring. Autistic children, whose disability is characterized by a fundamental inability to attribute mental states to others or to think of how they appear to others, are able to use a mirror to inspect their own bodies at the same age as normal children. And a number of experiments have shown that animals can use mirrors or closed-circuit television images to guide their hand motions. Baboons used a joystick to move a cursor so it followed a moving target on a computer screen, for example. Elephants were able to find hidden food by looking in a mirror, and chimpanzees could maneuver their hands to hidden targets using a television image that reversed the chimpanzees' hand movements. The common feature of all of these performances is that the animals had to adjust their hand (or trunk) motions according to some displaced and unfamiliar visual feedback. But the elephant did not need to know that that was *his* trunk in the mirror; all he had to learn was that when the trunk in the mirror touched the food in the mirror, his real trunk touched the real food—exactly as the baboon learned that when the cursor touched the moving dot, his hand was in the right place. The visual feedback taught the animals that when the hand or trunk or cursor on the screen was in position X, his real hand or trunk was in the corresponding position X-prime.

So being able to use information from a mirror to guide hand movements may be more akin to the guided, self-directed behaviors an animal shows when it scratches itself or maneuvers around an obstacle. It does not imply that the animal can "appreciate that its reflection resembles the way in which it is viewed" by another member of its species, says Heyes—and certainly not that it can contemplate its own mortality.

WHAT ANIMALS KNOW ABOUT OTHERS

Even if an animal knows nothing about its own thoughts or the thoughts of others, it may well know a great deal *about* others.

Evolution through natural selection is commonly thought of in terms of adapting to the external environment. A polar bear's claws and fur coat are clearly related to what a polar bear needs in order to make a living eating what it eats and living where it lives. But as Darwin himself recognized, a social animal may be subject to enormous selective pressures that come from the purely internal demands of the society it inhabits. The peacock's tail has nothing to do with adapting to the environment and everything to do with female mate choice. Male elk develop huge antlers for the sole purpose of threat displays and fighting with other elk in the competition for females. The advantage that these cumbersome physical appendages bring in reproductive success outweigh the enormous disadvantages they impose on the animals in coping with the environment at large—they cost energy and hinder the animal's ability to flee predators.

So, too, do many cognitive traits appear to have been selected purely for their value within an animal's society. Many group-dwelling animals appear to know a great deal about the members of their band, the blood relationships between them, and their standings in the social hierarchy. Interestingly, the cognitive hardware that has evolved to handle this sort of social knowledge often seems highly specialized. Male baboons appear to able to maintain long lists of dominance hierarchies among the other males in the group, but are unable to assess the relative amounts of water in a series of containers. Monkeys and apes that have been taught in the lab to learn sequencing tasks require considerable training to do so; in the wild they have no trouble spontaneously mastering mental representations of social relationships. As Dorothy Cheney and Robert Seyfarth found, vervet monkeys have developed a sophisticated social system of antipredator actions and vocalizations, but are remarkably stupid in attending to physical evidence from the environment—such as python tracks through the grass or carcasses of prey left by leopards—that would indicate the presence of predators. They are, as Cheney and Seyfarth put it, "primatologists who have spent too much time studying a single species." Good primatologists, but bad naturalists.

The compartmentalized and highly specialized nature of social cognitive abilities, and the fact that animals' social intelligence often seems to exceed nonsocial intelligence, tends to argue against interpreting an animal's knowledge of its society as evidence of a mental concept of self and other. Knowledge of knowledge ought to be more flexibly and universally applicable. In fact, possession of meta-representations is what fundamentally distinguishes a "universal" computing device from a special-purpose one. It is what distinguishes a programmable digital computer from a mechanical rocket guidance system. The ability to manipulate "thoughts" and not just to have them is what gives the computer its flexibility; in humans, language and mathematics perform a similar function at a higher level, allowing us to manipulate symbols and ideas independently of the special purposes our brain evolved to address. Our brains did not evolve so we could solve quadratic equations or play music or write books about the difference between human and animal minds.

The social knowledge that monkeys and apes possess is considerable, though it is difficult to say whether it is any greater than what would be found in chickens or wolves or horses or other group-dwelling animals where an ability to recognize individuals is important to survival in the group. Both chickens and long-tailed macaques have been shown to be able to recognize, as distinct individuals, the other members of their social group: The animals were first shown pairs of slides, one a picture of a group member and the other an unknown individual, and the animals were rewarded for picking the familiar face. Then they were presented a novel problem: Telling the difference between the same set of familiar and unfamiliar animals when the faces were shot at an angle, or when a picture of the entire animal was shown instead of a full-face shot. Regardless of the view, the animals did quite well in transferring their knowledge of individuals to these novel pictures of them.

Field studies of feral horses have likewise strongly suggested that members of a band can recognize, by color-markings, the several dozen other members of the group and know their relative social rank.

Experiments with monkeys both in the field and in the lab have underscored the sophisticated nature of their knowledge of social relationships. Particularly interesting were experiments with long-tailed macaques that were asked to match slides of mothers in their social group with slides of their offspring. A female macaque was shown slides of a mother along with two choices: that mother's own offspring, and another young monkey of the same age and sex. The monkey had to indicate its choice by pressing a button under each slide. After mastering the task of matching the mother-child pairs, the monkey was then shown a series of novel slides offering the same choice. She scored 90 percent correct. A second female macaque was given a slightly different task; she was shown pairs of slides containing either a mother-offspring pair or two unrelated members of the group. In the test phase, the wrong-answer slides always contained one adult female; the correct-answer, mother-offspring pairs included mothers with juvenile sons, mothers with adult daughters, and mothers with infant daughters. The female being tested scored fourteen out of fourteen correct.

The fact that the mother-offspring slides included some with grown offspring is particularly striking; the monkey seemed to be clearly recognizing their relationship rather than simply picking pictures that, say, contained young monkeys.

Experiments in the field are of course much more difficult to control, but some of Cheney's and Seyfarth's ingenious playback experiments are very persuasive in demonstrating the degree to which vervet monkeys recognize social relationships and not just individuals. The researchers taped individual monkey's calls and went to great lengths to ensure that the playback occurred under as controlled a condition as possible. The individual whose sounds were being played back had to be out of sight of the group being observed; the researchers also limited such experiments to one a day. In one experiment, they played back the scream of a two-year-old juvenile to a group of adult females. A significant number of the females responded by looking at the mother of the monkey whose voice they had just heard. One possible explanation is simply that these other females were reacting to the behavior of the mother, and not to the sound they had just heard—for, most of the time, the mother (unquestionably showing an ability to recognize the distinct sounds of its own offspring) would react by turning toward the loudspeaker. But even when the mother failed to react at all, the other females reacted by turning toward her. So there seems little doubt that the monkeys were not only able to recognize the sound of their own offspring but to recognize the sounds of others as well, and to match them up with their mothers.

Other signs that monkeys and apes recognize kin and status relationships come from field observations of dominance-related behaviors. Vervet monkeys in the wild preferentially seek to form bonds with higher-ranking individuals; they actively compete for the opportunity to groom the dominant members of the group. And both rhesus monkeys and vervets have been observed to selectively take out aggression on a close relative of an animal they have just been involved in a fight with. This sort of behavior by social analogy sometimes occurs even one additional step removed: A fight between A and B is often followed by an aggressive encounter between a relative of A and a relative of B.

Knowing who's who within one's social group is full of obvious practical value. Access to food and access to mates in particular is frequently determined by one's social standing, and avoidance of aggression is frequently determined by one's social savvy. Many animals know these social facts. But there is no particular evidence that they know what they know, or what others know.

WHAT ANIMALS KNOW ABOUT OTHERS' ABILITIES

When cheetah cubs are two-and-a-half to three-and-a-half months old, their mothers begin to refrain from killing prey themselves when they catch it. Instead, they capture the animal alive, carry it to the cubs, and let it go. The cubs typically run after the prey (usually rabbits or young gazelles) and knock it over repeatedly, though rarely succeeding in killing it themselves. After five to fifteen minutes, the mothers usually intervene and kill the prey.

By the time the cubs are about five months old, the mothers are releasing about a third of the prey they capture. As the cubs' skill in suffocating prey—or simply holding and eating prey alive—increases at around ten months, the mothers begin to release fewer animals. The mothers also modify their behavior depending on the type of prey; they are less likely to release mature gazelles, which the cubs are more likely to let escape.

Female house cats show a very similar behavior. When kittens start to be able to walk around, mother cats alter their normal hunting behavior; rather than eating captured prey at the site of the kill as they usually do, they start to bring the prey back to where the kittens are and eat it in front of them. As the kittens get older, the mother will start to release the captured prey they bring back and allow the kittens to play with it. If the kittens stop playing with the prey, the mother will start to play with the animal itself in an apparent (and usually successful) effort to renew the kittens' interest.

While cheetahs and cats are among the few species in which such behavior has been established by quantitative studies, anecdotes of similar efforts by parents to provide structured opportunities for offspring to learn abound. Otters have been observed to release captured prey before juveniles; meerkats will hold captured insects in their mouths and encourage their young to snatch bits away from them.

Birds of prey stage an even more elaborate series of training sessions for their young. Kestrels and sparrow hawks begin by leaving food at the nest for their young. Then, as the birds learn to fly, the adults begin to entice the young away from the nest; they will hold food in their talons or beak and make the young birds fly after them to be fed. Young ospreys subjected to this treatment would at first scream for food as usual, while the parents resolutely ignored them. Over the next several days, the young birds learned to follow their parents to a nearby rock, where they would be fed.

In the next stage, adults may drop food to young birds in flight, forcing them to catch it in midair for themselves. The ospreys, for example, would catch a fish, fly near one of their offspring, and drop it repeatedly. The parent would keep dropping and, if necessary, recatching the fish, until one of the young birds caught it. When one of the young failed to catch its own fish and alit on a rock where the other fledgling was eating its successful catch, the adult osprey landed, pushed the unsuccessful sibling off the rock, and forced him to keep trying to catch his own fish, which the adult again repeatedly dropped and caught until the young bird finally nabbed it himself. Some other birds of prey, much like the cats and cheetahs, drop small live birds for their young to go after in flight; peregrines have been seen to set up hunting practice for their young by swooping low over the ground to flush prey, which they leave for the young birds to try to catch for themselves.

What is particularly striking about many of these cases is how the adults will vary their "lessons" according to the response of their offspring. They are not merely following a rigid, preprogrammed set of actions. Once again we are tantalizingly close to evidence of an animal possessing a knowledge of the state of knowledge (or ignorance) of another. Perhaps the most compelling evidence is from studies of chimpanzees in natural settings. The chimpanzees of Taï National Park in the Ivory Coast, who we encountered in chapter 4 as exemplars of tool use and spatial maps, have been frequently observed to help their offspring learn how to use rocks as nutcracking hammers by leaving either piles of unopened nuts, or a hammer stone, or both on an "anvil stone" with a young chimpanzee who remains behind at the site. Adults normally consume the nuts they have placed on an anvil, and often carry a hammer with them rather than risk having it swiped by another chimpanzee. On about half of the occasions that mothers left a hammer behind with an infant, the infant would use it to open the nuts. This behavior is most often seen in mothers of chimpanzees that are at least three years old, the age at which they first begin to show interest in nuts. In many other cases a mother would allow her offspring to take the hammer away from her while she was still using it, and go search for a replacement for herself. And in two cases mothers seemed more directly to instruct their offspring in how to use a hammer properly. In one case the mother allowed her six-year-old son to take her hammer and most of the nuts she had collected. The young male placed a nut on the anvil, but before he opened it the mother approached and replaced the nut in a different position; the male then hit it with his hammer and ate the nut. In the second instance a mother rotated the hammer in the hand of her five-year-old daughter to a more effective position, then let her proceed to open the nut with it.

The fact that parental teaching is most commonly observed in carnivorous species—raptors, felines, primates—may argue that ecological, rather than cognitive, differences are the real driving factor here. Many equally "smart" species do not teach their offspring at all, even when common sense suggests there could be great survival value in their doing so. And a devil's advocate can always challenge the claim that parents who teach are aware of their offspring's mental states by pointing out that in every one of these instances the parents could do what they do by reacting to the overt *behavior* of their offspring alone. All mammals and birds appear to be able to notice and react to the changing needs of offspring as they grow, after all; teaching of hunting skills might be seen as merely one manifestation of this general cognitive ability, rather than an indicator of an exceptional cognitive ability to attribute mental states. All of the documented instances of "teaching" in animals involve a parent and its offspring.

Imitation, on the other hand, is seen in other social circumstances and might offer a more direct indicator of mental attribution—this time on the part of the "receiver" rather than the "sender." True imitation, of the sort that humans perform, requires seeing an action and then imagining what corresponding actions on one's own part would produce the same effect. But nailing down imitation is not easy. Animals in similar environments, after all, tend to behave similarly. Horses have traditionally been said to acquire certain bad habits or "stall vices" by imitating one another. These habits include pacing, or weaving back and forth, or cribbing—a particularly characteristic if odd behavior of horses in which they seize an object (such as the rim of a feed trough or a fence board) with their teeth, arch their necks, and gulp air. But the discovery that such behaviors are triggered in susceptible horses directly by the boredom and stress of prolonged confinement in a stall points to a more likely explanation: that horses do not catch these habits from one another, but from their shared environment.

Animal-rights advocates and New Age mystics out to prove that animals possess a higher consciousness have given considerable publicity to anecdotes reporting similar incidents in which new behaviors spread through an animal population. One tale of purported group consciousness in animals that has enjoyed a particularly long run is the so-called "Hundredth Monkey Phenomenon," which has been kicking around the New Age literature since 1979. The story is that primatologists in Japan had supplied a group of wild macaques on the Islet of Koshima with sweet potatoes. One young monkey discovered that sand could be removed from the potatoes by washing them in a stream, and this behavior began to spread through the population. Up to this point, the New Age tale agrees with a report that appeared in the journal *Pri-*

mates in 1965; the researchers called this phenomenon, which was first observed in 1953, an example of "precultural" transmission of behavior. But the New Age tale then took this mystical twist: the monkeys' newly acquired behavior spread slowly until the hundredth monkey had learned to wash sweet potatoes, when suddenly and spontaneously the behavior jumped into the consciousness of the entire population, and even leaped across the ocean to macaques on other islands and mainland Japan. The Hundredth Monkey Phenomenon is often cited as a sort of uplifting "you can make a difference" exhortation—if just enough committed activists come to see the evil of (fill in the blank: war, nuclear power, plastics, the oppression of women, etc.) then mass consciousness will take over and do the rest.

Needless to say, the Japanese primatologists reported nothing of the kind; the behavior spread slowly, through matrilineal lines, and recent analyses suggest it was not even the product of imitation but rather was the result of independent invention. In any case, the spread of potato washing was heavily conditioned by the fact that the food was being supplied artificially by human investigators.

Indeed, one of the most significant facts is that the behavior did *not* spread any faster even as more monkeys acquired it. If learning were actually taking place by imitation, the rate at which the behavior spread should have sped up as the number of "demonstrators" available to copy from grew.

The idea that each monkey that acquired potato-washing behavior invented it anew for himself (rather than the behavior having spread by imitation from a single creative genius) might seem farfetched. On the other hand, all of the monkeys at Koshima Islet experienced the same novel perturbation to their environment: people showed up and began giving them a food they had never experienced before and which was covered with sand. Other studies have shown that several monkey species have a natural propensity for food-washing; capuchin monkeys and crab-eating macaques both were able to learn food-washing very rapidly in captivity. On Koshima Islet, two other factors may have helped to create the misleading appearance of learning-by-imitation. One was that the caretaker on Koshima who fed the monkeys was observed by one visitor to give sweet potatoes only to those monkeys that washed them. In other words, monkeys were rewarded for doing it. The other factor is that the monkeys that washed sweet potatoes left potato skins scattered around at the bottom of the water. Baby monkeys following their mothers would pick up the skins out of the water, thus learning to associate potatoes with water from the start. In other words, it may not have been their mother's behavior they were learning, but merely a place associated with a certain food. They were thus primed to "reinvent" the behavior.

Similarly, the observed tendency among African chimpanzee bands that live in different areas to maintain different tool-use "traditions" may be explained in part by different environmental circumstances. The habit of poking twigs or blades of grass into holes appears to emerge as part of the normal play behavior in all infant chimpanzees. Poking these sticks into the same holes that adults are fishing termites out of would not, then, constitute an imitation of a behavior. In the natural course of things infants follow adults, they stick twigs into holes, and because the adults head for holes that have termites in them, those are holes the infants will naturally explore, too. This behavior is then reinforced when the young chimpanzees find that they get something to eat for their efforts. Adults tend to abandon their own probes outside termite nests where they have just used them, thus automatically creating further opportunities for infants to reinvent this specific tool-use behavior for themselves.

One reason that the chimpanzees in Gombe fish for termites with probes while the chimpanzees in Mahale do not is that (rather than its being a matter of a local "cultural tradition" that is passed down) the termites in Gombe taste good while the ones in Mahale don't. The latter secrete a distasteful chemical as a defensive tactic. Jane Goodall, while arguing that "it seems sensible to suppose" that tool-use behavior in chimpanzees is passed down through imitation, acknowledged that a simpler alternative is possible: "Undoubtedly, given the investigative and manipulative tendencies of the young chimpanzee and his ability to learn through trial and error, almost all of the feeding and tool using behaviors I have described could be invented anew by each individual, especially since the behavior of others in the group will serve to direct his attention to the relevant parts of the environment."

This process is sometimes called "local enhancement" or "stimulus enhancement," and it is extremely widespread in the animal kingdom. As researcher Michael Tomasello notes, animals are attracted to places where other members of their species are finding food; "this then places them in a position to learn something that they would not otherwise have learned, and what they learn is often identical to what the conspecifics are learning. . . . In local or stimulus enhancement nothing is actually learned from the behavior of others; the learner is simply attracted to a location or object."

Something that comes a step closer to imitation is a process that Tomasello calls "emulation learning"; again, the animal is not learning directly what behavior to reproduce, but learns by observation that a "change of state" in the environment is possible—though he still must rely on his own experimentation to discover what precisely it takes to make that change occur. Watching another chimpanzee use a rake to pull food toward a cage, for example, could teach another chimpanzee that food that is far from the cage can nonetheless move.

A number of experiments have confirmed that animals can learn about objects from other animals in this manner. Rats appear to learn to avoid certain foods by watching others (though rats may also urinate on noxious foods, which offers a more direct cue). In one especially interesting and rather amusing experiment, an octopus was placed in a tank with a red and a white ball, and rewarded for attacking the red one while ignoring the white one. When an "observer" octopus was allowed to watch the proceedings through a transparent partition and then given the same test, it too went for the red ball.

What is rare to nonexistent are instances in which animals learn to duplicate a complex behavior by watching another perform that behavior—in other words, true imitation. An ability to watch a sequence of movements by another and reproduce them oneself appears relatively late in human development, generally between twelve and eighteen months, by which time infants already have an ability to represent and manipulate ideas symbolically. The octopus has learned something about an object in the environment. But octopuses don't have to learn to attack things in the first place by watching other octopuses do it, just as chickadees do not have to learn to eat by watching other chickadees eating. Seeing a bird eating seed will attract other birds to the spot; they have learned something about the environment by watching others (where the food is), though not about behavior. Perhaps most disappointing of all to conventional wisdom is the total failure of experiments to support the idea that monkeys or apes are capable of imitation. Experiments have failed to show this, and what evidence has been reported, according to Bennett Galef, who conducted an extensive literature review, is almost entirely anecdotal. Monkey see, monkey don't necessarily do.

Significantly, when humans imitate others' behavior, they almost always perceive and understand the action in terms of *intentions;* they understand the action as "cleaning the window," for example, not "moving the hand in a circular motion on the surface of the window while holding a cloth," as Tomasello notes. True imitation may indeed depend on attributing intentions to others.

WHAT ANIMALS KNOW ABOUT OTHERS' MINDS

With such an accumulation of tantalizing hints of mental attribution by animals, some experimenters have tried to launch a more direct assault on the question of animals' "theory of mind." Many anecdotes are cited as evidence of theory of mind. For example, a female chimpanzee was seen to hold out its hand to another in an appeasement gesture, then attacked him when he drew near. The "theory of mind" interpretation of such a ruse is that the sneaky female knew that the male would be misled into approaching. The behaviorist spoilsport interpretation is that she had merely learned this trick by experience—having done it once in the past, and finding it achieved the desired result, she repeated it.

While some researchers have argued that eventually the sheer mass of such anecdotes outweighs alternative explanations, that is a contention that says more about the foibles of the human mind in drawing logical conclusions from probabilistic phenomena than it is good sense. "The plural of anecdote is not data," as one critical scientist noted. But we are all susceptible to this bias. A telling anecdote about interpretation of anecdotes: Matt Ridley, a science-writer colleague of mine, was an avid perpetrator of "crop circles" in his native England. These circles had mysteriously begun appearing in grain fields and seemed to defy scientific explanation. A perfect circle of flattened grain would spontaneously appear in a field. The public was enthralled with suggestions that these were the result of paranormal phenomena, alien spaceships, or previously unrecognized physical forces. After creating several of his own crop circles using a piece of rope attached to a stake, Ridley revealed in an article how easy it was to produce them and suggested the simplest explanation was that they were just hoaxes. This indeed shortly thereafter proved the case, when two extremely busy hoaxers came clean and revealed that for years they had been making hundreds of them. Ridley's prosaic explanation, however, invariably evoked this reaction: Oh, well, maybe hoaxes explain *some* of the crop circles, but of course it can't explain them all. His response, of course, was "Why not?" A hundred hoaxes does not turn a hoax into a paranormal phenomenon. The plural of anecdote is not data.

Formal experiments of knowledge attribution avoid the uncontrolled circumstances that bedevil even the most compelling anecdotes. But they must be very carefully designed if they are to distinguish between the sort of "mind reading" that is merely an evolved or learned tactic and the sort that truly reflects an ability to impute mental states to others. A horse may learn to look in the same direction as another horse that has just snorted and is standing still and alert with ears erect. But that may be nothing more than the product of experience. The horse may be merely "reasoning about observables," as C. M. Heyes put it in a brilliant critical essay about mental attribution by animals. Whenever another horse snorts and stares, the horse learns, there is something important (a predator, a person carrying a feed bucket) in the direction that that other horse is facing. This is the kind of "mind reading" inherent in many animal communication systems. An animal that makes a growl does not need to reason, "If I growl, I will sound like a big animal, and the animal that hears it will therefore think I am a big animal"; he is simply behaving in accordance with an evolved instinct that *takes advantage* of another evolved instinct by which an animal recognizes low-pitched sounds as ones to be avoided.

The other kind of "mind reading" is the true ability to impute intentions ("he is growling because he intends to bite me"). This is what Heyes calls "reasoning about mental states." In practice it can be devilishly hard to tell the two apart.

Reasoning about mental states automatically leads to an ability to perform feats of still higher-order reasoning. Daniel Dennett has suggested that one useful way to think about this is by categorizing degrees of intentionality. "Zeroth-order" intentionality reflects no intention; a dog growls because it is angry. First-order intentionality reflects an intention about one's own actions; a dog growls because it wants the other dog to get away from its food. But to have true second-order intentionality, an animal requires a theory of mind; it must have intentions about intentions (its own or another's)—the dog growls because it wants the other dog to understand it is mad. Once one understands that others have intentions, an endless number of higher orders of intentionality become possible; one can have intentions about intentions about intentions and so on. Dennett offers the example of our seeing the male character in the movie who is deliberately not smiling, and how we realize that he is not smiling because he does not want the beautiful girl to know that he knows she is dying for him to ask her to dance but does not want him to know it.

Part of the difficulty in designing experiments that can distinguish animals that possess a theory of mind from those that do not is that even what *appear* to be fairly high orders of intentional behavior can be the product of good old dumb evolution. One could interpret the unthinking intelligence behind the social signal of a growl as a case of third-order intentionality: the growler is trying to make the listener understand that it is trying to sound like a large animal (even though both know he is not a large animal). Furthermore, as we have already seen, a first-order feat of "reasoning about observables" can readily masquerade as a second-order feat of "reasoning about mental states."

One of the first experiments that claimed to demonstrate a theory of mind in chimpanzees involved giving chimpanzees the chance to fool human experimenters in the lab. The chimpanzee being tested got to watch while a person placed food in one of two concealed containers.

Then another person entered the room, and would open whichever of the two containers the chimpanzee pointed to. The people who opened the containers came in two different varieties, good guys and bad guys. The good guys wore green coats, acted in a friendly manner, and when they opened the container with food would share it equally with the chimpanzee. The bad guys wore white coats, sinister dark glasses, and black boots, and acted in a "brusque" manner toward the chimpanzee. The bad guys, when they opened the baited container, would eat all the food themselves and give the chimp none. Neither the good guys nor the bad guys actually knew which was the baited container, and they were instructed to pick whichever one they thought the chimpanzee was indicating to them. After 120 trials, the good guys consistently did a better job of choosing the right container than the bad guys did.

The researchers concluded that the chimpanzees had attributed different mental states to the two different sorts of trainers, and had accordingly adopted a policy of trying to deceive the bad guys. But of course the reasoning-about-observables explanation fits just as well: the chimpanzees could simply have learned from the color of the coats and the consequences of their actions what strategy to adopt. Their action was only "deception" if the chimpanzees knew in the first place that they were trying to communicate to another mind where the food was. If they merely were following a learned rule—pointing to the baited container gets you food in the green-coat case but results in the loss of the food in the white-coat case—they were reasoning about observables, not mental states.

Another experiment that is frequently cited as evidence of mental-state attribution founders on the same problem. In this experiment, a trainer (the "Knower") baited one of four containers. The chimpanzee could see the person, but a screen blocked its view of which container the food went into. Then a second trainer (the "Guesser"), who had been out of the room while the food was hidden, entered. The screen was removed, the Knower pointed to the baited container, and the Guesser pointed randomly to one of the three other containers. The chimpanzee was then allowed to search a single container of its choice and eat the food if it found it. After 150 trials, two out of four of the chimpanzees tested had learned to fairly consistently (60 to 70 percent of the time) search the container indicated by the Knower. Rhesus monkeys given the same treatment never learned to follow the Knower's advice, even after as much as four months of training. Four-year-old children learned to solve the problem in fewer than ten trials.

The researchers, led by Daniel Povinelli, concluded that the chimpanzees, but not the monkeys, were able to know that the person who saw the food hidden had a knowledge that differed from the person who had not seen it happen. But in a later paper, Povinelli conceded that the chimpanzees' performance could have been the result of rapid learning from observable events rather than a grasp of the knowledge states of others. The simple learned rule would be to choose the container pointed to by the person who was in the room when the food was hidden. No understanding of the knowledge state of that person is necessary to make such a correct deduction from the manifestly favorable results of following that rule. In fact, the chimpanzees' initial performance on a transfer experiment immediately following the "Knower" and "Guesser" experiment tends to confirm that they were learning a rule by trial and error rather than understanding something about mental states. In the transfer experiment, a third person hid the food in the presence of the chimpanzee, the Knower, *and* the Guesser while the Guesser wore a paper bag over his head. If the chimpanzees had learned from the initial experiment to make their choice on the basis of an understanding that the person who could see the food hidden was the person who possessed the knowledge, they ought to have immediately followed the Knower's choice in the transfer experiment. But instead, they performed at chance level for the first few trials of the transfer experiment. Only after making several wrong choices did they master the new rule—which presumably could be described as "go with the person who did not have the bag on his head."

The fact that apes master such tasks while monkeys do not may thus rest on factors other than knowledge attribution. It may be that monkeys have a harder time understanding where a person is pointing, or have a harder time recognizing individuals. It may be that they have a generally harder time forming learned associations. On the other hand, there is further negative evidence which strongly suggests that whatever the chimpanzees *are* doing, monkeys are *not* attributing mental states. Cheney and Seyfarth tried an experiment on Japanese macaques in which mothers were allowed to view a test area either alongside their offspring or with their offspring blocked off from view by a solid partition. Then with either the mother and child or just the mother watching, an experimenter either hid some food in a bin within the test area, or took up a menacing position crouching behind a screen and wearing a surgical mask and holding a net. The juvenile was then released into the test area. The idea was to see whether the mother would make a greater effort to alert her offspring either to the food or to the "predator" in the cases when the offspring had not been able to view the proceedings. Since macaques make distinctive food and predator calls, it seemed possible to monitor their internal "assumptions" about their offspring's state of knowledge. As it turned out, the monkeys called just as often to their "knowing" offspring as to their "ignorant" offspring.

Finally, a recent careful study by Povinelli has added additional negative evidence about even chimpanzees' ability to attribute mental states. In these experiments, young chimpanzees were first taught to request food from a human by extending their arms through whichever of two holes in a clear screen was directly in front of the person. Then two people stood behind the screen, one at each hole, and the chimpanzee had to choose which person to request food from. One of the persons would look at the chimpanzee (though without making direct eye contact, which could have been intimidating), while the second person was unable to see the chimpanzee for one reason or another—either because the person was blindfolded, or had his back to the chimp, or had a bucket over his head, or his hands over his eyes, or was looking off into a corner. The chimpanzees consistently failed to discriminate between the person who knew which hole the chimpanzee was reaching through and the person who did not. "The subjects showed no disposition for gesturing toward the experimenter who could see them," the researchers concluded. Significantly, the chimpanzees did follow the line of sight of the gaze of the experimenter who was staring off into space; yet "the use of the eyes of others" as a cue in this manner, the researchers concluded, "in no way uniquely implies a corresponding view into their mind." They can learn that where somebody looks matters without understanding that there is someone behind those eyes who is seeing.

The contrast with human children is instructive, for just as language studies show a desire on the part of children as young as nine months to direct parents' attention to objects as an end in itself, so more-direct studies of knowledge attribution show that young children can extract understanding from an appreciation of the mental states of others. For example, in one study of eighteen-to-twenty-four-month-old children, an adult says to the child, "Let's go find the 'modi'" (or some other nonsense word). Then he goes to a bucket and pulls out several objects, scowling each time, before pulling out one object and smiling. Nearly all the children, when asked which of the objects were "modis," knew that the last one was a "modi" and all the others were not.

Many animal researchers are fairly confident that more-sensitive experiments will eventually show that apes, at least, do possess some ability to attribute mental states. But the entire search has been a vivid reminder of the dangers of anthropocentrism. The things that apes are good at are the things they have evolved to do to survive in their particular ecological niche. And the things an animal is good at generally do not require three decades of ambiguous experiments to discover.

8

EVOLUTION AND RESPECT

HUMANS DO A LOT OF SCHEMING. WE HAVE ALREADY encountered the provocative argument that human consciousness was a response to the intense selective pressure upon a smart, group-dwelling animal to anticipate and counter the social stratagems of others. All group-dwelling animals need such talents to ward off the ever-present danger of violence, loss of social status, cuckoldry, infanticide, and all the other perils of living in close proximity to others of one's species. The unthinking intelligence of evolution has conferred such talents on many social animals; dogs have evolved an instinctive reaction to growls and whines as if they "knew" what they meant about the state of mind of the growler or whiner—even if they have no literal ability to impute motives or attribute states of mind to others. They may have no grasp beyond Dennett's "first-order" intentionality, but evolution has equipped them to react in ways that effectively reflect a higher-order understanding. They can "mind read" without knowing what a mind is.

In humans, the ability to more directly attribute motives and intentions via the symbolic manipulations of language conferred a far greater power for anticipation and staying a step ahead of one's opponent. Thus one theory for the evolution of consciousness and language is that it was an evolutionary "arms race" driven not by the needs to adapt to the world at large but rather to counter one's rivals within human society.

Yet, even if we assume that is true, the power of language is such that it can soar above whatever evolutionary purpose (or accident) drove its creation. Did we also evolve to do calculus? Or write poetry? Or lay bricks? Or paint landscapes? Or build computers? Is there a "module" of the brain for landscape gardening and one for automobile mechanics? Practitioners of so-called evolutionary psychology have tried very hard to argue that the human mind is indeed full of such specialized modules, and that behind each lies an evolutionary story of adaptation in the course of mankind's biological history. Maybe poetry reflects our ancestors' adaptations to finding rhythmical patterns in nature, and landscape gardening is an urge to recreate the savannas where *Homo sapiens* evolved. But to take everything people's minds do today and hunt for a neat, evolutionary "just-so" story to explain them seems absurd; for at some point the explanations just will not stretch any further. There are great similarities in human behavior across cultures, perhaps greater than we have always been willing to admit. No human culture exists that does not include among its defining characteristics war, institutionalized marriage, taboos against incest, the offering of food and drink as a gesture of hospitality, lying, humorous insults— and many other universals. Such universals surely point to a biological root to many human behaviors. Yet there are enough differences in human culture from place to place and generation to generation that at some point the search for evolutionary roots to behavior surely becomes futile. Evolutionary psychology cannot account for both slavery and its termination, or the generally accepted role of women in the nineteenth century and the generally accepted role of women in the twentieth century. Ideas do indeed have a life of their own. Language is a rocket that has escaped the gravitational pull of biological adaptation.

The ability to have thoughts about thoughts, which language gave us, is a discontinuous leap that took us, uniquely among all species, into a realm where ethical thought becomes possible. Guilt only has meaning where there are both intentions and a sense of the other. Empathy is only possible when we can sufficiently enter the mind of another to enter the experiences of another. Justice is only possible when we can conceive of moral concepts that transcend individual circumstances and behavior. "A man is not honest simply because he never had a chance to steal," goes a Jewish folk saying. From everything we know, that is not a notion that a chimpanzee could ever conceive of. The urge to better ourselves is not possible without the ability to see not only our own thoughts and behaviors but to imagine the thoughts and perceptions of others—so we can stand outside ourselves and see us as others see us.

The apparently universal religious impulse in man is closely intertwined with this. To imagine a god is the ultimate attribution of mental states: we attribute nothing short of omniscience to a being outside our own minds. This is the ultimate expression of Darwin's "utopian animal," for, in so doing, man has created the ultimate moral conscience; he has used his ability to attribute mental states as a check on his actions that transcends his unavoidably competitive relations with other members of the group. A literal "god" is an anthropomorphic container for this concept; but it is interesting that many people who find the literal notion of god uncongenial are just as likely as the more literal-minded to possess the sense of a "conscience" that seems to have an existence separate from their own thoughts and motives. Karen Armstrong, a former nun and the author of *A History of God,* tells in the introduction to her book that as she researched the history of the idea and experience of god, more than one "highly respected monotheist" of all three monotheistic religions told her "quietly but firmly" that God did not really exist, that God was in a very important sense the product of man's creative imagination—yet a creation that was the most important "reality" in the world.

Evolutionary psychologists can triumphantly point to the fact that for many, God is just another "person" to bargain with to maximize our self-interests; that God represents an extension of man's highly evolved tendency to use his intelligence as a tool for getting the best of social encounters; that from the start man has sought to propitiate an angry God through sacrifice and bargaining; that in prayer humans frequently reveal this tendency by offering to give up one thing in exchange for something they value more (please make my sick child well and I'll never miss church again). All of that is certainly true, but all of that ignores that as much a creature of biological determinism man remains, language has planted the seeds of occasional feats of transcendence. Man is a utopian animal, not a perfect animal.

ANIMALS, AS THEY ARE

The very fact of the human ability, through the indispensable medium of language, to attribute mental states and to grasp higher orders of intentionality has blinded us to the lower order of intentionality among animals. This is the curse of compulsive anthropomorphism, and it is by all appearances an incurable disease. But let us try for an instant to use our transcendent powers of linguistic reason to look beyond that, and see if there is any way we can more honestly describe what an animal mind is truly *like*. Introspections about *how* our unconscious and nonverbal thought processes work have proved notoriously unreliable. The fact is we just don't know—and even if we somehow did know, we have no means of describing cognitive processes that do not involve words. Albert Einstein wrote of thinking in visual imagery, but even that does not get us very far, for in a way he was describing only the end process of cognitive operations that themselves were inaccessible.

Yet introspections about what it is to *experience* unconscious and nonverbal thoughts might arguably offer more valid conclusions. Experimental evidence suggests that there is a great similarity between the unconscious thought processes of man and other animals. And so this may offer a clue to answer the question that in many ways is the one we really want answered: What is it like to experience the world as an animal does? When we look around, things catch our attention as they move or take on familiar shapes. We experience a "flash" of recognition when we see something we know; even before the word "squirrel" comes to the fore we have the sense of *knowing* that the thing we have seen belongs to a category that we recognize and associate certain properties with. We would jump if we saw a grizzly bear in our backyard well before our minds could supply the words *grizzly bear*, just as we react nonchalantly or perhaps with curiosity when we see a squirrel. Likewise we experience many emotions and sensations without the necessity to attach labels to them—pain, fear, hunger, thirst, surprise, pleasure, elation.

These are levels of sensations that it seems logical and justifiable to attribute to animals. Consciousness is quite another matter, though, for whether or not language causes consciousness, language is so intimately tied to consciousness that the two seem inseparable. The "monitor" that runs through our brains all the time we are awake is one that runs in language. The continual sense that we are aware of what is going on in a deliberate fashion is a sense that depends on words to give it shape and substance. Occasionally we may, while sitting on a beach, say, or a mountain top, have the sense that the monitor is turned off, that we are no longer particularly aware of ourselves, that the scene we are taking in is just there, that from time to time something moves and focuses our attention. More commonly we may take in something in a scene and only seconds—or sometimes even minutes—later be consciously aware of what our minds have seen and registered at some deeper level. All of those sensations may be a crude approximation of what it is "like" to be a dog or a horse or a monkey or a goldfish.

The premise of animal "rights" is that sentience is sentience, that an animal by virtue above all of its capacity to feel pain deserves equal consideration. But sentience is not sentience, and pain isn't even pain. Or perhaps, following Daniel Dennett's distinction, we should say that pain is not the same as suffering. "What is awful about losing your job, or your leg, or your reputation, or your loved one is not the suffering this event *causes* you, but the suffering this event *is*," Dennett writes. Our ability to have thoughts about our experiences turns emotions into something far greater and sometimes far worse than mere pain. The multiple shades of many emotions that our language expresses reveal the crucial importance of social context—of the thoughts we have *about* our experiences and the thoughts we have about those thoughts—in our perception of those emotions. Sadness, pity, sympathy, condolence, self-pity, ennui, woe, heartbreak, distress, worry, apprehension, dejection, grief, wistfulness, pensiveness, mournfulness, brooding, rue, regret, misery, despair—all express shades of the pain of sadness whose full meaning comes only from our ability to reflect on their meaning, not just their feeling. The horror of breaking a limb that we experience is not merely the pain; the pain is but the beginning of the suffering we feel as we worry and anticipate the consequences. Pity and condolence and sympathy are all shades of feeling that are manifestly defined by the social context, by the mental-state attribution to another that we are capable of. Consciousness is a wonderful gift and a wonderful curse that, all the evidence suggests, is not in the realm of the sentient experiences of other creatures.

By the same token an honest view of animal minds ought to lead us to a more profound respect for animals as unique beings in nature, worthy in their own right. The shallow and self-centered view that sees what is worthy in nature as that which resembles us seems vapid and petty by comparison. We try so hard to show that chimpanzees, or monkeys, or dogs, or cats, or rats, or chickens, or fish, or frogs are like us in their thoughts and feelings; in so doing we do nothing but denigrate what they really are. We define true intelligence and true feeling in human terms, and in so doing blind ourselves to the wonder of life's diversity that evolution has bequeathed earth. The intelligence that every species displays is wonderful enough in itself; it is folly and anthropomorphism of the worst kind to insist that to be truly wonderful it must be the same as ours.

It is always dangerous to try to draw moral lessons from the blindly amoral process of evolution. But if there is a lesson at all here, it is that all of the creatures that evolution has fashioned are remarkable in their own right. All have hit upon unique ways to make a living against all probability. And that is something to respect, and to treasure.

NOTES

Full references to the books and journal articles cited below in short form can be found in the bibliography.

INTRODUCTION

The story of Binti's "rescue" mission is in "Gorilla Saves Tot in Brookfield Zoo Pit," *Chicago Tribune,* 17 August 1996; "Zoo's New Top Banana," *Chicago Tribune,* 18 August 1996; "One Great Ape," *People,* 2 September 1996, 72. The subsequent discussions of maternal behavior in zoo-reared gorillas were reported in "Zookeeper Downplays 'Heroics' by Gorilla," *Columbus Dispatch,* 7 April 1997; "Animal Behavior: Binti Jua's Instincts Could Have Taken Over When Young Boy Fell into Zoo Enclosure," *Dayton Daily News,* 30 August 1996. The attitudinal survey of fundamentalist Christians is in Burghardt, "Animal Awareness," 905–6. For Aristotle and Darwin on human nature see Vidal and Vauclair, "Un animal politique," 35. "Maybe she felt closer to her dead offspring . . ." is in Masson, *When Elephants Weep,* 219. "Compulsive anthropomorphism" is discussed in Kennedy, *New Anthropomorphism,* 24–29. Searle's critique of the Turing test is summarized in Gardner, *Mind's New Science,* 17; Gray, "Consciousness." For the "cognitive revolution," see Vauclair, *Animal Cognition,* 7–11; Gardner, *Mind's New Science,* 32–40. Animal "Communicators" are reported on in "Getting in Touch With Your Felines," *Washington Post,* Style Section, 19 February 1997; "Mind Leap," *Utne Reader,* March-April 1998, 48–9. "We risk alienating ourselves psychologically" is in Savage-Rumbaugh, *Kanzi,* 253, 281. The review of the book on vervet monkeys is by R. W. Byrne in *The Sciences,* July 1990, 142–47. Goodall's foreword is in Rollin, *Unheeded Cry,* vii–ix. The rote nature of Kanzi's performance is discussed in Vauclair, *Animal Cognition,* 115; Terrace, "In the Beginning was the 'Name,'" 1013–14. "An unbreachable boundary between humans and nonhumans" is in Savage-Rumbaugh, *Kanzi,* 20, 252. "Behavioristic taboo" is in Griffin, *Animal Thinking,* vii. "A eulogy of animals" is in Thorndike, *Animal Intelligence,* 22–25. For radical behaviorists see Gray, "Consciousness." The fox that played dead is in Romanes, *Mental Evolution,* 314. "Ejective" method is in Romanes, *Mental Evolution,* 16–17, 22; Romanes,

Animal Intelligence, 420–22. Morgan's quoted criticisms are in Morgan, *Comparative Psychology,* 37, 53, 242–59, 290, 304. "No animal could learn to open a latched gate by accident" is in Thorndike, *Animal Intelligence,* 67. For discussions of Clever Hans, see Boysen and Capaldi, eds., *Numerical Competence,* 119; Budiansky, *Nature of Horses,* 165–66; Pfungst, *Clever Hans.* "Behaviorists began using the incident" is Gould quoted in Kluger, "Magna cum Critters." Capuchin monkey experiment is in Roitblat and Meyer, eds., *Cognitive Science,* 177. The performance of chimpanzees with sticks is in McFarland, ed., *Oxford Companion* 313, 468. For "Water bird" see Roitblat and Meyer, eds., *Cognitive Science,* 53.

CHAPTER 1. WHO IS THE SMARTEST OF THEM ALL?

Romanes' table of mental development is the frontispiece of Romanes, *Mental Evolution in Man.* For IQ tests see Gould, *Mismeasure of Man,* 199, 207–11. For relative brain size of animals see Calvin, *How Brains Think,* 11–12; McFarland, ed., *Oxford Companion,* 40–42. "If a goldfish was as intelligent as a chimpanzee" is Macphail, personal communication. The problem of motivation is discussed in Macphail, "Cognitive Function in Mammals," 280. Tests that found abilities previously thought to be the sole province of higher mammals are covered in Von Fersen et al., "Transitive Inference in Pigeons"; Devine, "Learning-Set Formation of Rhesus and Cebus Monkeys"; Roberts and Van Veldhuizen, "Spatial Memory in Pigeons"; Macphail, "Cognitive Function in Mammals," 282–85. For the null hypothesis see Macphail "Comparative Psychology of Intelligence," 648, 650; Macphail, "Cognitive Function in Mammals," 285. Specialized intelligence is discussed in Riley and Langley, "The Logic of Species Comparisons." For prepared and contraprepared tasks see Seligman, "Generality of the Laws of Learning." The grammatical-string experiments are described in Macphail, "Cognitive Function in Mammals," 288. Language as a discontinuity is in Gervet et al., "Evolution of Cognition," 42.

CHAPTER 2. THE SCIENCE OF HOW DO WE KNOW FOR SURE

Baboon deception is in Kummer, "Social Knowledge," 118. Firefly signaling is in Lloyd, "Firefly Communication," 113–16. For roosters' food calls see Marler et al., "Vocal Communication." For claims of intentional deception see Griffin, *Animal Minds,* 199; Vauclair, *Animal Cognition,* 134–35; Mills, "Unusual Suspects," 35. "Superstitious" dogs are in Holmes, *Farmer's Dog,* 82–83. "Ecological approach" is in Mace, "Strategy for Perceiving." Clever weeds are in Dennett, *Kinds of Minds,* 60–61, 154–55; Budiansky, *Covenant of the Wild,* 84–87. For deceptive strategies of prey and parasites see Mills, "Unusual Suspects"; McFarland, ed., *Oxford Companion,* 440. Flexibility as indicator of conscious intent is in Griffin, *Animal Thinking,* 24–25; Griffin, *Animal Minds,* 206–8. "It doesn't take a lot of neurons" is Caldwell quoted in Mills,

"Unusual Suspects." "Decision-making mechanisms . . . optimally adapted" is in Mc-Farland, "Goals, No-Goals," 46. "Permits our hypotheses to die in our stead" is in Popper quoted in Dennett, *Kinds of Minds*, 88. The "purposiveness" of evolution is discussed in Gallup, "Self-Awareness," 247; Kennedy, *New Anthropomorphism*, 86. For "mock anthropomorphism" see Dennett, *Kinds of Minds*, 27; Kennedy, *New Anthropomorphism*, 9, 87; Burghardt, "Animal Awareness." The Madagascar star orchid and dungfly examples are in Cockburn, *Evolutionary Ecology*, 90–92, 123–24. "It will often be a good guess" is in Kennedy, *New Anthropomorphism*, 93. Bushmen's "amazingly accurate" conclusions is in Fox, *Whistling Hunters*, 132. "Unless challenged to separate description . . ." is in Burghardt, "Animal Awareness." "Food calls" and courtship feeding are discussed in Marler et al., "Vocal Communication," 192; McFarland, ed., *Oxford Companion*, 112. For Moskowitz and Finkelstein see Rosten, *Joys of Yiddish*, 335. Horse vocalizations are discussed in Budiansky, *Nature of Horses*, 137–38. "Upping the ante" is in Vidal and Vauclair, "Un Animal Politique." The analogy of elephants as animal behaviorists is in Pinker, *Language Instinct*, 332–33. For anthropocentric versus ecological programs see Shettleworth, "Comparison in Comparative Cognition," 179–80. The Chomsky and Terrace quotes are in "Chimp Talk Debate: Is It Really Language," *New York Times*, Science Times Section, 6 June 1995.

CHAPTER 3. THE MIND'S SOFTWARE

For latent learning see Vauclair, *Animal Cognition*, 4–7. For the Turing machine see Newman, *World of Mathematics*, 2092–95; Gardner, *Mind's New Science*, 16–18. "Striving towards goals" is in Gardner, *Mind's New Science*, 20. For neural networks see Newman, *World of Mathematics*, 2089–92. "Invented a thinking machine" is in McCorduck, *Machines Who Think*, 116. "Symbols, rules, images" is in Gardner, *Mind's New Science*, 38–39. Cognitive disorders are described in Calvin, *How Brains Think*, 64; "What Is a Memory Made Of?" *U.S. News & World Report*, 18–25 August 1997, 71–73. For the Capgras delusion see Young, "Neuropsychology of Awareness." The number seven is discussed in Miller, "Magical Number Seven," 81; Calvin, *How Brains Think*, 92–93; Gardner, *Mind's New Science*, 89–91. Rotated-letter experiment is in Shepard and Metzler, "Mental Rotation." For localization of brain function see Posner et al., "Localization of Cognitive Operations"; Barinaga, "Visual System"; Nyberg et al., "General and Specific Brain Regions." The brain as "distributed" processor is discussed in Macphail, *Neuroscience of Animal Intelligence*, 19. The ELIZA dialogue is in Boden, *Artificial Intelligence*, 106–7. "Cog" is discussed in Dennett, *Kinds of Minds*, 15–16. For computer models of insects see Roitblat and Meyer, eds., *Comparative Approaches to Cognitive Science*, 41. The *E. coli* model is in Roitblat and Meyer, eds., *Comparative Approaches to Cognitive Science*, 159–61. The frog model is in Arbib and Cobas, "Schemas for Prey-Catching." The interaction of simple control loops with an environment is discussed in Churchland and Sejnowski, "Perspectives on

Cognitive Neurosciece," 745. For the cricket model see Webb, "A Cricket Robot." "When we say an animal perceives an intruder . . ." is in Prato Previde et al., "The Mind of Organisms," 91. For complex and hypercomplex cells see Gardner, *Mind's New Science*, 273–74; Maunsell and Newsome, "Visual Processing." The grid experiment is in Kosslyn, "Cognitive Neuroscience of Mental Imagery."

CHAPTER 4. USING THE OLD NOGGIN

Performance of goldfish and chimpanzees is discussed in Macphail, "Comparative Psychology of Intelligence." For horses learning to learn see McCall, "Learning Behavior in Horses," 76, 80. For conditional tasks see Vauclair, *Animal Cognition*, 12–14; McFarland, ed., *Oxford Companion*, 72, 311–16. Pigeons' perceptual difficulties are discussed in Macphail et al., "Relational Learning in Pigeons." Category learning by pigeons is in Herrnstein and de Villiers, "Fish as a Natural Category," 60–62; Vauclair, *Animal Cognition*, 15; Roberts and Mazmanian, "Concept Learning," 248. "Concept of an A" is in Morgan et al., "Pigeons Learn the Concept of an A." The kingfisher versus other-bird experiment is in Roberts and Mazmanian, "Concept Learning," 252–53. Baboon experiment is in Vauclair and Fagot, "Categorization of Characters." The family-resemblance model is discussed in Thompson, "Natural and Related Concepts"; Herrnstein and de Villiers, "Fish as a Natural Category," 80. Cat experiment is in Blake, "Cats Perceive Biological Motion." Timing of pigeons' and monkeys' responses is in Swartz, Chen, and Terrace, "Serial Learning," 401–4; Terrace, "List Learning"; D'Amato and Colombo, "Representation of Serial Order." How pigeons learn lists is in Terrace, "List Learning," 164, 166. Monkeys apply the transitive rule is in Harris and McGonigle, "Transitive Inference." Monkeys remember ordinal positions is in Chen, Swartz, and Terrace, "Knowledge of Ordinal Position." Birds know the number of eggs is in Seibt, "Are Animals Attuned to Number?," 597. For laboratory studies of numerical judgment see Honig, "Numerosity as a Dimension of Stimulus Control," 61–63, 80; Davis, "Discrimination of the Number Three"; Davis and Pérusse, "Numerical Competence in Animals," 570–72. Subitizing and rhythmic representations of numbers are discussed in Matsuzawa, "Use of Numbers by a Chimpanzee"; Davis and Prusse, "Numerical Competence in Animals," 563–64. Pigeons' relative numerousness judgments are discussed in Honig, "Numerosity as a Dimension of Stimulus Control," 80–82. Monkeys "count down" is in Rumbaugh and Washburn, "Counting by Chimpanzees," 102–6; Boysen et al., "Processing of Ordinality." Sheba's performance is in Boysen and Berntson, "Numerical Competence in a Chimpanzee." "Heroic effort" to train Sheba is in Boysen and Capaldi, eds., *Development of Numerical Competence*, 39. "Absolute numerosity a human invention" is in Davis, "Numerical Competence in Animals," 109–10.

A note on calculating Solomon's performance: The chance of hitting the right answer on each try on each problem in $1/3 = 33$ percent. So, for example, the chance that Solomon will make no wrong answers at all—that is, he will guess right on all five problems the first time he encounters each one—is $(.33)^5 = 0.4$ percent. The chance he will hit the right answer on the second try on any given problem is $(2/3)(1/3) = 22$ percent. So his chance of getting at most one wrong answer in total is the sum of the probability of getting none wrong at all, plus the probabilities of getting one wrong on problem 1 and zero wrong on all the rest, one wrong on problem 2 and zero on all the rest, and so on: $0.4 + (.22)(.33)(.33)(.33)(.33) + (.33)(.22)(.33)(.33)(.33) +$ etc. This works out to about 1.7 percent. Continuing this process up to 20 permissible wrong answers yields a probability of 95 percent. Note that this assumes chimpanzees do not learn from an unreinforced wrong answer not to try that same wrong answer again; they simply guess at random each time until they get it right, which means they may try the same wrong answer more than once. If, on the other hand, they do learn to rule out wrong answers, on average they will make only 5 errors out of the 82 trials (94 percent correct), and it becomes a complete certainty that by the time they make 10 errors they will have solved all five problems (having tried and ruled out all possible wrong answers).

CHAPTER 5. MAPS, TOOLS, AND NESTS

Clark's nutcrackers and marsh tits are discussed in Griffin, *Animal Minds*, 45–47. Construction of cognitive maps is in Gallistel and Cramer, "Computations on Metric Maps," 211–12. Reaction of baboons to novel objects is in Vauclair, *Animal Cognition*, 69–70. Baboons and mental rotations is in Vauclair, Fagot, and Hopkins, "Rotation of Mental Images in Baboons." For shortcuts by bees see Dyer, "Bees Acquire Route-Based Memories"; Bennett, "Do Animals Have Cognitive Maps?" 221. The wind-tunnel experiment with bees is in Kirchner and Braun, "Dancing Honey Bees." For shortcuts by dogs see Chapuis and Varlet, "Short Cuts by Dogs." Critique of chimpanzees' "traveling salesman" problem solution is in Bennett, "Do Animals Have Cognitive Maps," 222. How monkeys find food caches is in Cramer, "Computations on Metric Cognitive Maps"; Gallistel and Cramer, "Computations on Metric Maps," 215. For hammer transports by chimpanzees see Boesch and Boesch, "Mental Map in Wild Chimpanzees." Rat hippocampus experiments are discussed in O'Keefe and Burgess, "Geometrical Determinants of Place Fields"; McNaughton, "Cognitive Cartography." Disoriented rats and toddlers is in Hermer and Spelke, "Geometric Process for Spatial Reorientation." Reorientation with visual cues is discussed in O'Keefe, "Cognitive Maps in Infants?" For pigeon navigation see McFarland, ed., *Oxford Companion*, 401–4. For solitary wasps see Tinbergen, *Study of Instinct*, 9; McFarland, ed., *Oxford Companion*, 311. For tool-use in insects and finches see Griffin, *Animal Minds*, 102–3; McFarland, ed., *Oxford Companion*, 576. For a discussion of nest building see McFarland, ed., *Oxford*

Companion, 408–11, 579. Tool use by chimpanzees is in Boesch and Boesch, "Tool Use and Tool Making." For tool-use experiments with apes and monkeys see Visalberghi et al., "Performance in a Tool-Using Task." Capuchin monkey experiments are in Visalberghi and Trinca, "Tool Use in Capuchin Monkeys"; Gibson and Ingold, eds., *Tools, Language, and Cognition*, 139. For ecological niche and tool use see Chevalier-Skolnikoff and Liska, "Tool Use by Elephants." For tool use and language development see Calvin, *How Brains Think*, 96–97; Gibson and Ingold, eds., *Tool Use, Language, and Cognition*, 193, 241; Calvin, "Emergence of Intelligence."

CHAPTER 6. SPEAK!

For ground squirrels and alarm calls see Morton and Page, *Animal Talk*, 56, 218–21. Whines and growls are explained in Morton, "On the Occurrence of Motivational-Structural Rules." The evolution of honest signaling is discussed in Guilford and Dawkins, "Receiver Psychology," 9–10. The fallacy of "deception" is discussed in Guilford and Dawkins, "Receiver Psychology," 10; Dawkins and Krebs, "Animal Signals." "GOP" is in Griffin, *Animal Minds*, 155. Playback experiments with birds are in Morton and Page, *Animal Talk*, 179. For arms races and dialects in bird songs see Morton and Page, *Animal Talk*, 190–91; 200–201. The role of random errors in dialects is in Williams and Slater, "Simulation of Song Learning." Vervet monkey alarm calls are discussed in Cheney and Seyfarth, "Précis," 140. For mistakes by young vervet monkeys see Snowdon, "Sounds of Silence"; Owings, "Calls as Labels." Mental-state attribution by monkeys is in Tomasello, "Cognitive Ethology"; Cheney and Seyfarth, "Précis," 142. "Nim's signing was prompted by his teachers" is in Terrace, *Nim*, vi; Terrace et al., "Can an Ape Create a Sentence?" "Unspontaneous and imitative nature" of Nim's signs is in Terrace, "Evidence for Sign Language." "It's exactly like the ASL sign for 'give'!" is quoted in Pinker, *Language Instinct*, 337–38. "I know what it's like" is in "Chimp Talk Debate: Is It Really Language?" *New York Times*, Science Times Section, 6 June 1995. The popular book for children is Michel, "The Story of Nim." Kanzi, Sherman, Austin, and Lana are discussed in Terrace, "In the Beginning Was the 'Name,'" 1011–14, 1024–25; Savage-Rumbaugh et al., "Spontaneous Symbol Acquisition." "Dolphins also account for these two features" is quoted in "Can Animals Think," *Time*, 22 March 1993. For a discussion of grammar versus rote sequences see Vauclair, *Animal Cognition*, 111; "Thinking About Dolphins," *National Wildlife*, April/May 1994, 5–9. For toddlers' use of words see Smillie, "Rethinking Piaget's Theory of Infancy," 292; Baron-Cohen, "How Monkeys Do Things." Joint attention between mothers and infants is discussed in Terrace, "In the Beginning Was the 'Name,'" 1019–20. Apes rarely manipulate objects to engage infants' attention is in Vauclair and Vidal, "Discontinuities in the Mind." The naming task with Sherman and Austin is described in Savage-Rumbaugh et al., "Can a Chimpanzee Make a Statement?" 479. "The child would be thought of as disturbed" is in "Chimp Talk Debate: Is It Really Language?" *New York Times*, Science Times Section, 6 June 1995.

CHAPTER 7. ANIMAL CONSCIOUSNESS

Self-awareness and meta-representations are discussed in Asendorpf and Baudon-nière, "Self-awareness and Other-Awareness," 89; Schull and Smith, "Knowing Thy-self, Knowing the Other." Adaptive value of self-awareness is in Humphrey, "Nature's Psychologists." Rats reporting their own behaviors is in Beninger et al., "Ability of Rats to Discriminate Their Own Behaviors." Bailout experiments are described in Schull and Smith, "Knowing Thyself, Knowing the Other"; Smith et al., "Uncertain Response in the Bottlenosed Dolphin"; "Probing Primate Thoughts: Questions Arise About the Mental Lives of Apes and Monkeys," *Science News*, 20 January 1996. Gallup's mirror-recognition test is in Gallup, "Chimpanzees: Self-Recognition." Critiques of mirror recognition are found in Swartz and Evans, "Not All Chim-panzees Show Self-Recognition," 493; Heyes, "Reflections on Self-Recognition," 911. Stumptailed macaques and mirrors are discussed in Anderson, "Monkeys with Mirrors," 85. Control-group behavior is discussed in Heyes, "Reflections on Self-Recognition," 911–13. Cotton-top tamarins are in Hauser et al., "Self-Recognition in Primates," 10813. Use of visual feedback is in Vauclair and Fagot, "Manual and Hemispheric Specialization"; Vauclair, *Animal Cognition* 143–44. "The way in which it is viewed" is in Heyes, "Reflections on Self-Recognition," 918. Vervet monkeys' social knowledge is in Cheney and Seyfarth, "Précis," 137–38. Teaching by cheetahs and cats is in Caro and Hauser, "Teaching in Nonhuman Animals," 156–59. Teach-ing by raptors and chimpanzees is in Caro and Hauser, "Teaching in Nonhuman An-imals," 163–64, 170. For the Hundredth Monkey "phenomenon" see Possel and Amundson, "Senior Researcher Comments on Hundredth Monkey"; Kawai, "Newly Aquired Precultural Behavior." How the Koshima Islet macaques actually learned to wash potatoes is analyzed in Galef, "Question of Animal Culture," 162–66. Imita-tion in chimpanzees is discussed in Galef, "Question of Animal Culture," 166, 171. For social enhancement see Tomasello, "Do Apes Ape?" 321. For imitation in hu-mans see Tomasello, "Do Apes Ape?" 323. Reasoning about observables and mental states is in Heyes, "Attribution of Mental States," 180. Orders of intentionality are discussed in Dennett, *Kinds of Minds*, 119–21. Mental-state attribution experiments in chimpanzees are in Povinelli et al., "Do Rhesus Monkeys Atttibute Knowledge?"; Povinelli and Eddy, "What Chimpanzees Know About Seeing," 22–23; Heyes, "At-tribution of Mental States," 182–83; Heyes, "Cues, Convergence," 242. Lack of mental-state attribution by monkeys is in Cheney and Seyfarth, "Attending to Behav-iour Versus Knowledge." The bucket-over-the-head experiment is in Povinelli and

Eddy, "What Chimpanzees Know About Seeing," 50, 105–7. "Modi" experiment is in Povinelli and Eddy, "What Chimpanzees Know About Seeing," 166.

CHAPTER 8. EVOLUTION AND RESPECT

"Highly respected monotheist" is in Armstrong, *History of God*, xix–xx. For pain versus suffering see Dennett, *Kinds of Minds*, 164–67.

BIBLIOGRAPHY

Anderson, James R. "Monkeys with Mirrors: Some Questions for Primate Psychology," *International Journal of Primatology* 5 (1984): 81–98.

Arbib, Michael A., and Alberto Cobas. "Schemas for Prey-Catching in Frog and Toad." In *From Animals to Animats*, Jean-Arcady Meyer and Stewart W. Wilson, eds. Cambridge, MA: MIT Press, 1991.

Armstrong, Karen. *A History of God.* New York: Knopf, 1993.

Asendorpf, Jens B., and Pierre-Marie Baudonnière. "Self-Awareness and Other-Awareness: Mirror Self-Recognition and Synchronic Imitation Among Unfamiliar Peers," *Developmental Psychology* 29 (1993):88–95.

Barinaga, Marcia. "Visual System Provides Clues to How the Brain Perceives," *Science* 275 (1997): 1583–85.

Baron-Cohen, Simon. "How Monkeys Do Things with 'Words,'" *Brain and Behavioral Sciences* 15 (1992): 148–49.

Beninger, Richard J., Stephen B. Kendall, and C. H. Vanverwolf. "The Ability of Rats to Discriminate Their Own Behaviours," *Canadian Journal of Psychology* 28 (1974): 79–91.

Bennett, Andrew T. D. "Do Animals Have Cognitive Maps?" *Journal of Experimental Biology* 199 (1996): 219–24.

Blake, R. "Cats Perceive Biological Motion," *Psychological Science* 4 (1993): 54–57.

Boden, Margaret. *Artificial Intelligence and Natural Man.* New York: Basic Books, 1977.

Boesch, Christophe, and Hedwige Boesch. "Mental Map in Wild Chimpanzees: An Analysis of Hammer Transports for Nut Cracking," *Primates* 25 (1984): 160–70.

————. "Tool Use and Tool Making in Wild Chimpanzees," *Folia Primatologica* 54 (1990): 86–99.

Boysen, Sarah T., and Gary G. Berntson. "Numerical Competence in a Chimpanzee *(Pan troglodytes)*," *Journal of Comparative Psychology* 103 (1989): 23–31.

Boysen, Sarah T., and E. John Capaldi, eds. *The Development of Numerical Competence: Human and Animal Models.* Hillsdale, NJ: Lawrence Erlbaum Associates, 1993.

Boysen, Sarah T. et al. "Processing of Ordinality and Transitivity by Chimpanzees *(Pan troglodytes)," Journal of Comparative Psychology* 107 (1993): 208–15.

Budiansky, Stephen. *The Covenant of the Wild: Why Animals Chose Domestication.* New York: Morrow, 1992. Reprint. Leesburg, VA: Terrapin Press, 1995.

———. *The Nature of Horses: Exploring Equine Evolution, Intelligence, and Behavior.* New York: Free Press, 1997.

Burghardt, Gordon M. "Animal Awareness. Current Perceptions and Historical Perspective," *American Psychologist,* August 1985, 905–19.

Calvin, William H. "The Emergence of Intelligence," *Scientific American,* October 1994, 101–7.

———. *How Brains Think.* New York: Basic Books, 1996.

Caro, T. M., and M. D. Hauser. "Is There Teaching in Nonhuman Animals?" *Quarterly Review of Biology* 67 (June 1992): 151–74.

Chapuis, Nicole, and Christian Varlet. "Short Cuts by Dogs in Natural Surroundings," *Quarterly Journal of Experimental Psychology* 39B (1987): 49–64.

Chen, Shaofu, Karyl B. Swartz, and H. S. Terrace. "Knowledge of the Ordinal Position of List Items in Rhesus Monkeys," *Psychological Science* 8 (1997): 80–86.

Cheney, Dorothy L., and Robert M. Seyfarth. "Attending to Behaviour versus Attending to Knowledge: Examining Monkeys Attribution of Mental States," *Animal Behaviour* 40 (1990): 742–53.

———. "Précis of *How Monkeys See the World, " Behavioral and Brain Sciences* 15 (1992): 135–82.

Chevalier-Skolnikoff, Suzanne, and Jo Liska. "Tool Use by Wild and Captive Elephants," *Animal Behaviour* 46 (1993): 209–19.

Churchland, Patricia S., and Terrence J. Sejnowski. "Perspectives on Cognitive Neuroscience," *Science* 242 (1988): 741–45.

Cockburn, Andrew. *An Introduction to Evolutionary Ecology.* Oxford: Blackwell, 1991.

Cramer, Audrey E. "Computations on Metric Cognitive Maps: How Vervet Monkeys Solve the Traveling Salesman Problem." Ph.D. diss., University of California, Los Angeles, 1995.

D'Amato, M. R., and M. Colombo. "Representation of Serial Order in Monkeys *(Cebus apella)," Journal of Experimental Psychology: Animal Behavior Processes* 14 (1988): 131–39.

Davis, Hank. "Discrimination of the Number Three by a Raccoon *(Procyon lotor),"* Animal Learning and Behavior* 12 (1984): 409–13.

———. "Numerical Competence in Animals: Life Beyond Clever Hans." In *The Development of Numerical Competence: Human and Animal Models.,* Sarah T. Boysen and E. John Capaldi, eds. Hillsdale, NJ: Lawrence Erlbaum Associates, 1993.

Davis, Hank, and Rachelle Pérusse. "Numerical Competence in Animals: Definitional Issues, Current Evidence, and a New Research Agenda," *Behavioral and Brain Sciences* 11 (1988): 561–615.

Dawkins, Richard, and John R. Krebs. "Animal Signals: Information or Manipulation?" In *Behavioural Ecology: An Evolutionary Approach,* John R. Krebs and Nicholas B. Davies, eds. Oxford: Blackwell, 1978.

Dennet, Daniel C. *Kinds of Minds.* New York: Basic Books, 1996.

Devine, J. V. "Stimulus Attributes and Training Procedures in Learning-Set Formation of Rhesus and Cebus Monkeys," *Journal of Comparative and Physiological Psychology* 73 (1970): 62–67.

Dugatkin, Lee Alan, and Anne Barrett Clark. "Of Monkeys, Mechanisms, and the Moduar Mind," *Behavioral and Brain Sciences* 15 (1992): 153–54.

Dyer, Fred C. "Bees Acquire Route-Based Memories but Not Cognitive Maps in a Familiar Landscape," *Animal Behaviour* 41 (1991): 239–46.

Fox, Michael W. *The Whistling Hunters. Field Studies of the Asiatic Wild Dog.* Albany, N.Y.: SUNY Press, 1984.

Galef, Bennet G., Jr. "The Question of Animal Culture," *Human Nature* 3 (1992): 157–78.

Gallistel, C. R., and Audrey E. Cramer. "Computations on Metric Maps in Mammals: Getting Oriented and Choosing a Multi-Destination Route," *Journal of Experimental Biology* 199 (1996): 211–17.

Gallup, Gordon G. "Chimpanzees: Self-Recognition," *Science* 167 (1970): 86–87.

―――. "Self-Awareness and the Emergence of Mind in Primates," *American Journal of Primatology* 2 (1982): 237–48.

Gardner, Howard. *The Mind's New Science: A History of the Cognitive Revolution.* New York: Basic Books, 1987.

Gervet Jacques et al. "Some Prerequisites for a Study of the Evolution of Cognition in the Animal Kingdom," *Acta Biotheoretica* 44 (1996): 37–57.

Gibson, Kathleen R., and Tim Ingold, eds. *Tools, Language, and Cognition in Human Evolution.* Cambridge: Cambridge University Press, 1993.

Gould, Stephen Jay. *The Mismeasure of Man.* New York: Norton, 1981.

―――. "Evolution: The Pleasures of Pluralism," *New York Review of Books,* 26 June 1997, 47–52.

Gray, Jeffrey. "Consciousness on the Scientific Agenda," *Nature* 358 (1992): 277.

Griffin, Donald R. *Animal Thinking.* Cambridge, MA: Harvard University Press, 1984.

―――. *Animal Minds.* Chicago: Chicago University Press, 1992.

Guilford, Tim, and Marian Stamp Dawkins. "Receiver Psychology and the Evolution of Animal Signals," *Animal Behaviour* 42 (1992): 1–14.

Harris, M., and Brendan McGonigle. "Modelling Transitive Inference," *Quarterly Journal of Experimental Psychology* 47B (1944): 319–48.

Hauser, M. D. et. al. "Self-Recognition in Primates: Phylogeny and the Salience of Species-Typical Features," *Proceedings of the National Academy of Sciences* 92 (1995): 10811–14.

Hermer, Linda, and Elizabeth S. Spelke. "A Geometric Process for Spatial Reorientation in Young Children," *Nature* 370 (1994): 57–59.

Herrnstein, R. J., and Peter A. de Villiers. "Fish as a Natural Category for People and Pigeons," *The Psychology of Learning and Motivation* 14 (1980): 59–95.

Heyes, C. M. "Anecdotes, Training, Trapping and Triangulating: Do Animals Attribute Mental States?" *Animal Behaviour* 46 (1993): 177–88.

———. "Reflections on Self-Recognition in Primates," *Animal Behaviour* 47 (1994): 909–19.

———. "Cues, Convergence, and a Curmudgeon: A Reply to Povinelli," *Animal Behaviour* 48 (1994): 242–44.

Holmes, John. *The Farmer's Dog.* Revised edition. London: Popular Dogs, 1975.

Honig, Werner K. "Numerosity as a Dimension of Stimulus Control." In *The Development of Numerical Competence: Human and Animal Models.*, Sarah T. Boysen and E. John Capaldi, eds. Hillsdale, N.J.: Lawrence Erlbaum Associates, 1993.

Humphrey, Nicholas K. "The Social Function of Intellect." In *Growing Points in Ethology,* P. P. G. Bateson and R. A. Hinde, eds. Cambridge: Cambridge University Press, 1976.

Kawai, Masao. "Newly Acquired Precultural Behavior of the Natural Troop of Monkeys on Koshima Islet," *Primates* 6 (1965): 1–30.

Kennedy, J. S. *The New Anthropomorphism.* Cambridge: Cambridge University Press, 1992.

Kirchner, Wolfgang H., and Ulrich Braun. "Dancing Honey Bees Indicate the Location of Food Sources Using Path Integration Rather than Cognitive Maps," *Animal Behaviour* 48 (1994): 1437–41.

Kluger, Jeffrey. "Magna cum Critters," *Discover,* January 1995.

Kosslyn, Stephen M. "Aspects of a Cognitive Neuroscience of Mental Imagery," *Science* 240 (1988): 1621–26.

Kummer, Hans. "Social Knowledge in Free-ranging Primates." In *Animal Mind— Human Mind,* Donald R. Griffin, ed. Berlin: Springer-Verlag, 1982.

Lloyd, James E. "Firefly Communication and Deception." In *Deception, Perspectives on Human and Nonhuman Deceit,* R. W. Mitchell and N. S. Thompson, eds. Albany, NY: SUNY Press, 1986.

Mace, William M. "James J. Gibson's Strategy of Perceiving: Ask Not What's Inside Your Head, but What Your Head's Inside Of." In *Perceiving, Acting, and Knowing: Toward an Ecological Psychology,* Robert Shaw and John Bransford, eds. Hillsdale, NJ: Lawrence Erlbaum Associates, 1977.

Macphail, Euan M. "The Comparative Psychology of Intelligence," *Behavioral and Brain Sciences* 10 (1987): 645–95.

———. *The Neuroscience of Animal Intelligence.* New York: Columbia University Press, 1993.

———. "Cognitive Function in Mammals: The Evolutionary Perspective," *Cognitive Brain Research* 3 (1996): 279–90.

Macphail, Euan M. et al. "Relational Learning in Pigeons: The Role of Perceptual

Processes in Between-Key Recognition of Complex Stimuli," *Animal Learning and Behavior* 23 (1995): 83–92.

Marler, Peter et al. "Vocal Communication in the Domestic Chicken II," *Animal Behaviour* 34 (1986):194–98.

Masson, Jeffrey Moussaieff, and Susan McCarthy. *When Elephants Weep: The Emotional Lives of Animals.* New York: Delacorte Press, 1995.

Matsuzawa, Tetsuro. "Use of Numbers by a Chimpanzee," *Nature* 315 (1985): 57–59.

Maunsell, J. H. R., and W. T. Newsome. "Visual Processing in the Monkey Extrastriate Cortex," *Annual Review of Neuroscience* 10 (1987): 363–401.

McCall, C. A. "A Review of Learning Behavior in Horses and its Application in Horse Training," *Journal of Animal Science* 68 (1990): 75–81.

McCorduck, Pamela. *Machines Who Think.* San Francisco: W.H. Freeman, 1979.

McFarland, David. "Goals, No-Goals, and Own Goals." In *Goals, No-Goals, and Own Goals,* Alan Montefiore and Denis Noble, eds. London: Unwin Hymn, 1989.

McFarland, David, ed. *Oxford Companion to Animal Behavior.* Oxford: Oxford University Press, 1987.

McNaughton, Bruce. "Cognitive Cartography," *Nature* 381 (1996): 368–69.

Michel, Anna. "The Story of Nim, the Chimp Who Learned Language." New York: Knopf, 1980.

Miller, George A. "The Magical Number Seven, Plus or Minus Two: Some Limits on Our Capacity for Processing Information," *Psychological Review* 63 (1956): 81–97.

Mills, Cynthia. "Unusual Suspects," *The Sciences,* July/August 1997, 32–36.

Morgan, C. Lloyd. *Introduction to Comparative Psychology.* London: Walter Scott, 1894.

Morgan, M. J. et al. "Pigeons Learn the Concept of an 'A'," *Perception* 5 (1976): 57–66.

Morton, Eugene S. "On the Occurrence and Significance of Motivational-Structural Rules in Some Bird and Mammal Sounds," *American Naturalist* III (1977): 855–69.

Morton, Eugene S., and Jake Page. *Animal Talk.* New York: Random House, 1992.

Newman, James R. *The World of Mathematics.* New York: Simon and Schuster, 1956.

Nyberg, Lars, et al. "General and Specific Brain Regions Involved in Encoding and Retrieval of Events: What, Where, and When," *Proceedings of the National Academy of Sciences* 93 (1996): 11280–85.

O'Keefe, John. "Cognitive Maps in Infants?" *Nature* 370 (1994): 19–20.

O'Keefe, John, and Neil Burgess. "Geometrical Determinants of the Place Fields of Hippocampal Neurons," *Nature* 381 (1996): 425–28.

Owings, Donald H. "Calls as Labels: An Intriguing Theme, but One with Limitations," *Behavioral and Brain Sciences* 15 (1992): 162–63.

Pfungst, Oskar. *Clever Hans: The Horse of Mr. Von Osten.* 1911.

Pinker, Steven. *The Language Instinct.* New York: Morrow, 1994.

Posner, Michael I et al. "Localization of Cognitive Operations in the Human Brain," *Science* 240 (1988): 1627–31.

Possel, Markus, and Ron Amundson. "Senior Researcher Comments on the Hundredth Monkey Phenomenon in Japan," *Skeptical Inquirer*, May-June 1996, 51.

Povinelli, Daniel J., and Timothy Eddy. "What Young Chimpanzees Know About Seeing," *Monographs of the Society for Research in Child Development* 61, no. 3 (1996).

Povinelli, Daniel J., Kathleen A. Parks, and Melinda A. Novak. "Do Rhesus Monkeys *(Macaca mulatta)* Attribute Knowledge and Ignorance to Others?" *Journal of Comparative Psychology* 105 (1991): 318–25.

Povinelli, Daniel J. et al. "Self-Recognition in Chimpanzees *(Pan troglodytes)*: Distribution, Ontogeny, and Patterns of Emergence," *Journal of Comparative Psychology* 107 (1993): 347–72.

Prato Previde, Emanuela et al. "The Mind of Organisms: Some Issues About Animal Cognition," *International Journal of Comparative Psychology* 6 (1992): 79–100.

Riley, Donald A., and Cynthia M. Langley. "The Logic of Species Comparisons," *Psychological Science* 4 (1993): 185–89.

Roberts, William A., and Dwight S. Mazmanian, "Concept Learning at Different Levels of Abstraction by Pigeons, Monkeys, and People" *Journal of Experimental Psychology: Animal Behavior Processes* 14 (1988): 247–60.

Roberts, William A., and N. Van Veldhuizen. "Spatial Memory in Pigeons in the Radial Arm Maze," *Journal of Experimental Psychology: Animal Behavior Processes* 11 (1985): 214–59.

Roitblat, Herbert L., and Jean-Arcady Meyer, eds. *Comparative Approaches to Cognitive Science.* Cambridge, MA: MIT Press, 1996.

Rollin, Bernard E. *The Unheeded Cry: Animal Consciousness, Animal Pain, and Science.* Oxford: Oxford University Press, 1989.

Romanes, George J. *Animal Intelligence.* 1882.

———. *Mental Evolution in Animals.* 1883.

———. *Mental Evolution in Man.* 1888.

Rosten, Leo. *The Joys of Yiddish.* New York: Pocket Books, 1970.

Rumbaugh, Duane M., and David A. Washburn. "Counting by Chimpanzees and Ordinality Judgments by Macaques in Video-Formatted Tasks." In *The Development of Numerical Competence: Human and Animal Models.*, Sarah T. Boysen and E. John Capaldi, eds. Hillsdale, NJ: Lawrence Erlbaum Associates, 1993.

Savage-Rumbaugh, Sue, and Roger Lewin. *Kanzi: The Ape at the Brink of the Human Mind.* New York: John Wiley & Sons, 1994.

Savage-Rumbaugh, Sue et al. "Can a Chimpanzee Make a Statement?" *Journal of Experimental Psychology: General* 112 (1983): 457–87.

———"Spontaneous Symbol Acquisition and Communicative Use by Pygmy Chimpanzees *(Pan paniscus)*," *Journal of Experimental Psychology: General* 115 (1986): 211–35.

Schull, Jonathan, and J. David Smith. "Knowing Thyself, Knowing the Other: They're Not the Same," *Behavioral and Brain Science* 15 (1992): 166–67.

Seibt, Uta. "Are Animals Naturally Attuned to Number?" *Behavioral and Brain Sciences* 11 (1988): 597–98.

Seligman, Martin E. P. "On the Generality of the Laws of Learning," *Psychological Review* 77 (1970): 406–18.

Shepard, Roger, and Jacqueline Metzler. "Mental Rotation of Three-dimensional Objects," *Science* 171 (1971): 701–3.

Shettleworth, Sara J. "Where Is the Comparison in Comparative Cognition?" *Psychological Science* 4 (1993): 179–84.

Smillie, D. "Rethinking Piaget's Theory of Infancy," *Human Development* 25 (1982): 282–94.

Smith, J. David et al. "The Uncertain Response in the Bottlenose Dolphin *(Tursiops truncatus)*," *Journal of Experimental Psychology: General* 124 (1995): 391–408.

Snowdon, Charles T. "The Sounds of Silence," *Behavioral and Brain Sciences* 15 (1992): 167–68.

Swartz, Karyl B., and Siûn Evans. "Not All Chimpanzees *(Pan troglodytes)* Show Self-Recognition," *Primates* 32 (1991): 483–96.

Swartz, Karyl B., Shaofu Chen, and H. S. Terrace. "Serial Learning by Rhesus Monkeys: I. Acquisition and Retention of Multiple Four-Item Lists," *Journal of Experimental Psychology: Animal Behavior Processes* 17 (1991): 396–410.

Terrace, H. S. *Nim.* New York: Knopf, 1979.

———. "Evidence for Sign Language in Apes: What the Ape Signed or How Well the Ape Was Loved?" *Contemporary Psychology* 27 (1982): 67–68.

———. "In the Beginning Was the 'Name,'" *American Psychologist*, September 1985, 1011–28.

———. "The Phylogeny and Ontogeny of Serial Memory: List Learning by Pigeons and Monkeys," *Psychological Science* 4 (1993): 162–69.

Terrace, H. S. Shaofu Chen, and Vikram Jaswal. "Recall of Three-Item Sequences by Pigeons," *Animal Learning and Behavior* 24 (1996): 193–205.

Terrace, H. S., and Brendan McGonigle. "Memory and Representation of Serial Order by Children, Monkeys, and Pigeons," *Current Directions in Psychological Science* 3, no. 6 (1994): 180–85.

Terrace, H. S., et al. "Can an Ape Create a Sentence?" *Science* 206 (1979): 891–902.

Thompson, Roger K. R. "Natural and Relational Concepts in Animals." In *Comparative Approaches to Cognitive Science*, Herbert L. Roitblat and Jean-Arcady Meyer, eds. Cambridge, MA: MIT Press, 1996.

Thorndike, Edward L. *Animal Intelligence.* 1911.

Tinbergen, Nickolaas. *The Study of Instinct.* Reprint, Oxford: Clarendon Press, 1989.

Tomasello, Michael. "Cognitive Ethology Comes of Age," *Behavioral and Brain Sciences* 15 (1992): 168–69.

————. "Do Apes Ape?" In *Social Learning in Animals: The Roots of Culture,* C. M. Heyes and Bennet G. Galef, Jr., eds. New York: Academic Press, 1996.

Vauclair, Jaques. *Animal Cognition: An Introduction to Modern Comparative Psychology.* Cambridge, MA: Harvard University Press, 1996.

————. *L'intelligence de l'animal.* Paris: Éditions du Seuil, 1992.

————. "Mental States in Animals: Cognitive Ethology," *Trends in Cognitive Sciences* I (1997): 35–39.

Vauclair, Jacques, and Joël Fogot. "Manual and Hemisphere Specialization in the Manipulation of a Joystick by Baboons," *Behavioral Neuroscience* 107 (1993): 210–14.

————. "Categorization of Alphanumeric Characters by Guinea Baboons: Within- and Between-Class Stimulus Discrimination," *Current Psychology of Cognition* 15 (1996): 449–62.

Vauclair, Jacques, Joël Fagot, and William D. Hopkins. "Rotation of Mental Images in Baboons when the Visual Input Is Directed to the Left Cerebral Hemisphere," *Psychological Science* 4 (1993): 99–103.

Vauclair, Jacques, and Jean-Marie Vidal. "Discontinuities in the Mind Between Animals and Humans." Paper presented at the Berder Conference on Cognition and Evolution, 10 March 1994.

Vidal, Jean-Marie, and Jacques Vauclair, "Un animal politique autre qu'humain" *Epokhi* 6 (1996): 35–55.

Visalberghi, Elisabetta, and Loredana Trinca. "Tool Use in Capuchin Monkeys: Distinguishing Between Performing and Understanding," *Primates* 30 (1989): 511–21.

Visalberghi, Elisabetta et al. "Performance in a Tool-Using Task by Common Chimpanzees *(Pan troglodytes),* Bonobos *(Pan paniscus),* an Orangutan *(Pongo pygmaeus),* and Capuchin Monkeys *(Cebus apella),*" *Journal of Comparative Psychology* 109 (1995): 52–60.

Von Fersen, L. et al. "Transitive Inference in Pigeons," *Journal of Experimental Psychology: Animal Behavior Processes* 17 (1991): 334–41.

Webb, Barbara. "A Cricket Robot," *Scientific American,* December 1996, 94–99.

Williams, James, and P. J. B. Slater. "Simulation Studies of Song Learning in Birds." In *From Animals to Animats,* Jean-Arcady Meyer and Stewart W. Wilson, eds. Cambridge, MA.: MIT Press, 1991.

Young, Andrew. "The Neuropsychology of Awareness." In *Consciousness in Philosophy and Cognitive Neuroscience,* Antii Revinsuo and Matti Kamppinen, eds. Hillsdale, NJ: Lawrence Erlbaum Associates, 1994.

INDEX

Printed in the United States
By Bookmasters